Listening to Beauty

Listening to Beauty

Rhetorics of Science in Sea and Sound

MEGAN POOLE

The University of Chicago Press
Chicago and London

The University of Chicago Press, Chicago 60637
The University of Chicago Press, Ltd., London
© 2025 by The University of Chicago
All rights reserved. No part of this book may be used or reproduced in any manner whatsoever without written permission, except in the case of brief quotations in critical articles and reviews. For more information, contact the University of Chicago Press, 1427 E. 60th St., Chicago, IL 60637.
Published 2025

34 33 32 31 30 29 28 27 26 25   1 2 3 4 5

ISBN-13: 978-0-226-83867-0 (cloth)
ISBN-13: 978-0-226-84287-5 (paper)
ISBN-13: 978-0-226-84288-2 (e-book)
DOI: https://doi.org/10.7208/chicago/9780226842882.001.0001

Library of Congress Cataloging-in-Publication Data

Names: Poole, Megan, author.
Title: Listening to beauty : rhetorics of science in sea and sound / Megan Poole.
Description: Chicago : The University of Chicago Press, 2025. |
  Includes bibliographical references and index.
Identifiers: LCCN 2024059216 | ISBN 9780226838670 (cloth) |
  ISBN 9780226842875 (paperback) | ISBN 9780226842882 (ebook)
Subjects: LCSH: Animal sounds. | Marine animals. | Bioacoustics. |
  Animal communication. | Listening (Philosophy) | Aesthetics.
Classification: LCC QL765 .P66 2025 | DDC 591.59/4—dc23/eng/20250115
LC record available at https://lccn.loc.gov/2024059216

For Andrew

Contents

List of Figures ix
Introduction 1

1 Nature's Punctum 17
2 Punctive Listening 33
3 Extractive Listening 58
4 Emergent Listening 83
5 Swallowed by Beauty 108
Conclusion 137

Acknowledgments 143
Notes 147
Bibliography 169
Index 183

Figures

1. Diffraction patterns   21
2. Spectrograms of whale songs on the cover of *Science*   37
3. Whale watching (and listening) featuring Katy Payne and Annie Lewandowski   42
4. Listening to elephants featuring Katy Payne   62
5. *Siren—Listening to Another Species on Earth*   84
6. Male king bird of paradise   119
7. Outer tail feathers of a male king bird of paradise   120
8. *Habronattus* jumping spider   125

Introduction

The first time I remember witnessing beauty—really witnessing it, remembering what was once beautiful rather than having the scene right there before my eyes—I was standing amid an oak grove chenier along coastal Louisiana. Trees standing and branching and bending and swaying for hundreds of years now stood shriveled, white-gray limbs bare against a bright summer sky. What was the first wooded resting stop for tropical birds migrating from Central and South America gives testimony to an uncertain future following a series of destructive hurricanes fed by climate change. Standing there, staring up into this tangle of broken limbs, I search for words to explain what it means to witness this scene, which some have called a "ghost forest," a forest of dead trees starved of leaves and bark by salt water, wondering what I should have done differently the last time I saw this grove in its fullness. Because how do you describe this ecological rarity that exists only along the far western side of the Mississippi Delta—this chenier, or "oak" in Cajun French—to those who were not present before my home parish succumbed to the rising Gulf of Mexico? To be honest, I may have never asked this question had it not been for so many hurricanes. The beauty of places like Oak Grove came into relief only in its fading away.

Down there, in our rush to save memories, wildlife, landmarks, and history from erasure, we are all learning to specialize in witnessing. We are attending to how time warps, never neatly stretching out behind and before us but folding backward and forward on itself. We are learning that our testimonies resonate, calling what was once home into presence so that our connection to the place can be experienced again.[1] In many ways, this book is about how observers become witnesses; how bodies hold lessons from the land; how beauty, moving in us and through us, asks us to stop and listen.

More pointedly, though, I am concerned with how bodies can come to such conclusions. It would not be controversial to say that bodies are able to know through lived experience, but that remark explains too much away. *Listening to Beauty* considers how our bodies learn through lived experience, how our natural environments challenge what we think we know, how we understand lessons from nonhuman kin, even when those lessons are coded in signs and symbols that resist discursive ciphering.

Perhaps it is surprising to turn here, at this question of how the wordless can be witnessed, to the tradition of rhetoric. Often recognized as the sending and receiving of arguments—if not, specifically, finding the "available means of persuasion" in political speeches or popular texts—rhetoric is well known as communication acts that carry overt demands for attention or calls to action. Yet this notion depicts rhetoric primarily as the end product of communication. Rhetoric's substance, however, is in how what we think, know, feel, and say emerge from our environments, in what has to happen—discursively, sure, but also affectively, bodily, aesthetically—for our learned ways of being and knowing to be disrupted. Rhetoric allows us to understand how bodies receive, hold, and issue messages, even those messages spoken not in words. I study rhetoric, then, to understand how this oak grove calls me to attend as much to an uncertain future and my place in it as to how the former beauty of the place that shaped me ushers in such knowledge.

Of course, my depiction of Oak Grove's desolation is not quite fair—this "ghost forest" displays its own kind of beauty. Witnessing that beauty is, admittedly, tricky because there is no one thing that can be recalled or pointed to as the definitive account of that place. No one thing that "killed" the forest, no way to bear witness and move on. Oak Grove is beautiful today in its rotting: detritus serving as new habitat for shorebirds benefiting from sand blown too far inland by the hurricanes. This mutability—never static, fixed, or predictable—marks the kind of beauty I am after in this book. Beauty here is as much in the manicured as the chaotic, in the quotidian as the transcendent, in the memory of a lush oak grove as the ghost forest caught me unaware. Or beauty is in the interplay of what are too often considered dichotomous poles. Rather than resolving meanings, beauty amplifies contradictions. Otherwise, I may not still be wondering how all the memories we stand to lose if that chenier never recovers are not just caught up in *our bodies*—human bodies—but also in the spatial memory of the whooping crane and the root networks of coastal prairie grasses. When I stand on that chenier, more than an appreciation for natural beauty is occurring. The beauty in this natural environment is raising questions.[2] I may consider these questions rhetorically, scientifically, aesthetically, or diffractively, taking what resonates across each approach, but the point is that I take these questions—and our ability to receive these questions from nonhuman kin—seriously.[3] The beautiful may not exactly help us understand how nonhuman animals and environments convey their knowledge, but beauty may just teach us the prerequisite required for opening ourselves to a kind of aesthetic knowledge that reaches us sensorially and affectively, rather than discursively. That prerequisite is listening.

To better understand what is happening in this oak grove, or in other moments where our experiences in natural environments cannot be made sense of through words and yet overwhelm us with a profound knowing, I have interviewed experts who specialize in nonhuman communication.[4] In *Listening to Beauty*, you will hear from eight researchers in evolutionary biology, bioacoustics, and marine ecology to analyze what has to happen, bodily and sensorily, to feel elephants' infrasonic rumbles or attend to the songs of the humpback whale.[5] Interested in how these researchers detect, hold, and interpret sensory messages from their natural environments, I asked them how questions emerge during fieldwork, how we might linger in the uncertainty of scientific studies, how to handle interference that comes from subjective interpretation, how to interact with nonhuman kin, how to make sense of beauty. My questions, at first, perplexed them; these researchers were accustomed to queries about their methods, their tools, their disciplinary interventions—not their relationship to beauty. As each interviewee searched for words to describe their ways of knowing in the field and being with their subjects of study, each turned, eventually and without prompting, to listening.

Here is my claim: Beauty can be constitutive of listening, or, at least, beauty can be integral to our learning to listen. Listening cannot occur without sustained attention and a careful discipline or comportment of the body—beauty is one such experience that convinces, if not teaches, us to linger and attend, to comport otherwise. To be clear, we are not talking about *that* kind of beauty, those cultural imperatives that teach us what to like or what is beautiful, that branch of aesthetics that privileges certain individuals over others, those with enough taste and prestige to judge the beautiful. Indeed, we will not bother with questions of "what makes something beautiful" or even with evolutionary questions about "why beauty exists," although those questions circle around my conversations with these researchers. Indeed, you will encounter the beautiful plumage of birds, the melodies of marine mammals, and the emotional care of elephants, but our primary focus will center on what happens when the human biologists attending to those colors, songs, and rumbles reach the limits of discourse, why they turn—rhetorically—to the idea of "beauty" in those encounters, and how their natural environments reach forth to grab their attention in the first place. *Listening to Beauty* interrogates those wordless moments of encountering beauty and argues that the surge of aesthesis in these encounters serves as an epistemological resource to scientific invention, one integral to logic, data, objectivity.

At the onset, allow me to get ahead of something: the impossibility of writing the wordless, applying form to the unsayable. The best I can do is ask you not to take only my word for it. Sit down with these researchers along with me, listen to their stories, draw your own conclusions. We do not often see this side of scientific studies—the side that lingers in awe and wonder—but these stories may just teach us all how better to attend to our natural environments, how to listen to the beauty presently before us, and how to be empathetic witnesses to all we do not yet know. Allowing science's stories to hold as much, if not more, weight than the theories and citations, bring us closer to beauty than scholarly books often do.

That is why I have written this book in a way that does not presume a value of the work's theoretical origins over the grounded theory stemming from scientists' stories. You need not read the rest of this introduction or the entirety of the first chapter on "nature's punctum" to learn from the biologists in this book. With the idea of nature's punctum, I mean an appeal from our natural environments that disrupts logical, disciplinary ways of understanding and brings us to know sensorially, affectively, and more than discursively. Nature's punctum calls our attention to how beautiful phenomena move us—what elements register with our experiences, how meaning emerges from places we cannot articulate, how what compels our attention is broader than that of which we can make sense. The first chapter engages with histories, philosophies, and rhetorics of science, offering an extended definition of "nature's punctum," but the biologists interviewed in this book provide and extend their own sort of definition. Their approach to listening in the field and a new appreciation for beauty are available if you wish to skip ahead to the whales in chapter 2, the elephants in chapter 3, the dolphins in chapter 4, and the birds in chapter 5. If you are curious about my theorizing before then, we will walk through the fraught history of aesthetics in relation to science's "modest witness," redefining beauty not in matters of taste or value but as a rupture in the sensible that ushers in more than discursive ways of knowing.

Sensing Beauty

We are turning to "beauty" because each researcher in this book has talked, in their own way, about their studies as "beautiful." Consider, as we will later in more detail, one of the first instances in which "beauty" was used by biologists in the 1970s to describe the sounds of the humpback whale. Whales were not just sounding but singing, revealed Katy Payne, a musician turned bioacoustics expert. Or consider how Payne unearthed the second of her scientific findings: researchers were unable to decipher much about elephant communication until Payne sat in silence with elephants for hours and felt bodily vibrations as if the lowest note on an organ had been struck. For Payne, this feeling in her chest suggested that elephant communication was infrasonic. Payne's methods illustrate how the rhetorical work of engaging our environments is not always so rational. In those moments when encounters with nonhuman kin bring us to the limits of discourse and disciplinary logic, the biologists in this study show that revering beauty may tune us in to listening, to adopting a bodily comportment that promotes inventive, creative paths for advancing knowledge, even scientific knowledge.

In the public sphere, too, beauty is having a moment. When I began writing this book in 2018, the film *Won't You Be My Neighbor?*, featuring children's television host Mr. Rogers—the voice who reminded viewers each morning that "it's a beautiful day in the neighborhood"—became the top-grossing biographical documentary of all time. Ocean Vuong's novel *On Earth We're Briefly Gorgeous* was named one of the top ten books of 2019 by *The Washington Post* and spent six weeks on *The New York Times* Best Seller list. In Athens, the Hellenic National Archaeological Museum curated a special exhibit in 2018 titled *The Countless Aspects of Beauty* to consider the aesthetics of the mundane in early antiquity. In Spain, an exhibit called *The Beautiful Brain* presented viewers with the art of neuroscientific imaging, and then traveled for years to museums across the globe. Such depictions of beauty were not so much concerned with how humans adorned themselves or with highbrow matters of taste, as is often the case when we talk about beauty. Instead, these examples found beauty in the everyday and banal, in the language of an immigrant mother, in the terra-cotta jar used to carry water, in the divergent branching of the neuron. Amid all too frequent news of violence and extinctions, these were humanistic attempts to figure out where we lost a sense for the beautiful and what we might do to witness beauty—in all its forms and becomings—once again.

At the same time, apart from searching for beauty aesthetically, scientists and science writers began interrogating the biological need for beauty. In 2018 ornithologist Richard O. Prum's *The Evolution of Beauty* was a finalist for the Pulitzer Prize and argued that by refusing to take Darwin's ideas about beauty in natural environments seriously, biologists have misunderstood the foundations of evolutionary theory. Four years later, science writer Ed Yong's *An Immense World* considered all the beauty often undetected before us by examining insect and animal senses. In *Becoming Wild*, ecologist Carl Safina investigated the beautiful plumage, songs, and lifelong partnerships of scarlet macaws to hypothesize that "beauty makes us love what it takes to live."[6] These more scientific investigations challenged the idea that beauty in natural environments could be attributed merely to instinct and acknowledged the presence of messy processes like culture, communication, the ability to be moved, swayed, and caught up in feeling among our nonhuman kin.

These returns to beauty in creative and scientific endeavors signaled an important shift in the history of aesthetics. For too long, philosopher of art and art critic Arthur C. Danto articulates, the study of the beautiful *was* the study of aesthetics, with beauty considered as the highest aesthetic value, with art created for the purpose of representing beauty, and with the experience of beauty as evidence that a given piece of art "worked," not to mention that the beholder possessed the education and experience necessary to judge that artwork as beautiful. Once modern and postmodern artwork exposed beauty as an "option for art and not a necessary condition," the beautiful became a problem with the white male gaze or a trivialized refrain of appreciation, an element only in the eye of the beholder.[7] I take the attempts above, then, as efforts to reclaim beauty from an elitist, racist past stemming from ideas of human exceptionalism.

Rather than merely an aesthetic ideal, beauty highlights the resources of a different way of knowing, an "aesthetic intelligibility" that is affective, more than discursive, and deeply revelatory. So argues philosopher of art Robert B. Pippin, who finds in Hegelian and Kantian aesthetics how experiencing the beautiful brought viewers to recognize their duality as, at once, "naturally embodied objects in the world and, without consistency, practically free, responsible agents."[8] How Kant attended to the difference between beauty in human artwork and beauty in natural environments indicates as much for Pippin, who argues "that nature was beautiful, and that we were able to appreciate that beauty in a way that was not merely hedonic or standardly cognitive, all meant something [for Kant], [and] was of some decisive significance."[9] On that much, we agree: studying how beauty emerges from natural environments—attending not just to what human artists create—can lead philosophers to different conclusions about aesthetics.

What conclusions can be reached and how best to study the emergence of beauty are where our perspectives diverge, however, and this is, in many ways, the starting point of this project. That is, Pippin concludes that "aesthetic intelligibility is a distinct sensible-affective modality of intelligibility" and that beauty relays "sensible-affective markers of truth, or at least truthfulness, genuineness, in some concrete form of self-understanding."[10] And

INTRODUCTION 7

he concludes as much through analyses of Western canonical artwork. In this schema, beauty remains an end of artwork, a way to experience self-understanding, a form of enlightenment for the human. *Listening to Beauty* likewise admits an "aesthetic intelligibility," but so as not to replicate the ideological harms of the Western artwork tradition, this study departs entirely from notions of "art." More important here is beauty not as an ideal end or marker of value; rather, beauty is a conduit to ways of knowing that can only be sensed. That is, we will come to understand beauty, to use the words of Black studies scholar Christina Sharpe, as method.

Not just any method, beauty is a *rhetorical* method that turns us to the resources of sensing, to ways of knowing aesthetically. By invoking "aesthetics" here and throughout the rest of the book, I mean sense perception in its broadest, richest valences and in its earliest instantiation in ancient Greek etymology—not the criticism of art objects, but aesthetics as the study of how we experience and sense our environments. For ancient philosophers, rhetoric scholar Debra Hawhee explains, *aisthēsis* translated to sense perception or sensation more generally.[11] Similarly, philosopher Katya Mandoki defines aesthesis as "receptivity, openness to the environment, the sentient and sensorial activity in any scale."[12] Aesthetics, then, is the study of how we sense our environments, how we open ourselves to our surroundings in order to receive affect, feeling, emotion, maybe even beauty, as well as how we make sense of what happens in those encounters.

More importantly, this study considers not just the *how* but also the *who* of aesthetics, to whom we are in relation as we sense our environments. At least for Aristotle, sensation is the mode of engagement that binds together human and nonhuman kin.[13] This "shared arena of sensation" promotes interspecies connections and interactions in which feeling, knowing—if not also meaning making—can be mutual, or common among a myriad of relations.[14] If the history of aesthetics frequently returns to ideas of individualized self-understanding or matters of taste, this study of aesthetics queries what all is shared in sensory encounters across species. Further, if rhetoric involves the study of how bodies hold more than discursive messages received from our environments, then aesthetics is part and parcel of that study.[15]

We are sensing our environments at every moment, of course, not often aware of every ounce of sensory stimuli presently available for perception. When quotidian, mundane sensing turns into heightened perception, when the body perceives its limits for symbolic knowing and turns to aesthetic ways of knowing—that is the experience of beauty. Or awe, wonder, horror, the sublime, but beauty is the experience here under query. We might say that "beauty" is the word that these biologists turn to when they reach the limits of the explainable, a catch-all term that signals a deep sense of appreciation or humility before their subjects of study. But remember that beauty is more than a rhetorical term. Beauty is a method that draws the scientist to consider, study, know otherwise.

Beauty is thus a resource in the stage of rhetorical invention, that preparatory stage of intellectual creativity and receptivity prior to analysis in which thinkers gather what is known about a subject in preparation to make a hypothesis or argument. Typically considered a discursive process, rhetorical invention is also an aesthetic process, affectively laden and sense-inflected, marked by a "wordlessness" of thought.[16] Think of rhetorical invention as working on a spectrum, with one pole as the affective milieu in which meaning is able to be sensed but not articulated through words, and the other pole as the discursive realm in which established terms are accessible to convey an argument or establish a truth claim.[17] *Listening to Beauty* is concerned with what happens on the "wordless," affective side of the spectrum and how studying that side of rhetorical invention in the sciences allows insight into the capacious sensory, material, and aesthetic resources that challenge scientific knowledge and broaden established ways of knowing. Caught up as it is in the wordless, this study of the emergence of beauty requires a heavy dose of listening.

Listening

Rhetoric scholars are well-versed in nonrational appeals, observing when more than discursive matter gives way to meaning or, more pointedly, considering how matter offers its own sort of appeals.[18] Yet how such appeals influence and undergird scientific research—those approaches led by objective methods of inquiry that produce discursive, replicable facts—is less pronounced, given a tendency to focus more on how scientists persuade rather than how they are themselves persuaded by their nonhuman subjects of study. Studying how scientists are persuaded by impositions in the field requires considering the researcher as a listener more than a speaker.[19] Listening is "the other side of language," according to rhetoric scholar Daniel M. Gross, or the side of language that clues us in on how bodies are moved through nonlinguistic registers like feeling, affect, and emotion.[20] Asking how someone receives this other side of language also gets at questions that are prior to how one knows; these questions are about how one holds themselves in the world in order to know. Attuning to listening allows us to examine how these biologists receive and understand meaning that appeals through more than discursive registers as well as distancing themselves from the ideas and paradigms of their disciplines in order to allow nonhuman kin and natural environments to speak, as much as possible, for themselves.

Perhaps most importantly, bringing attention to how these scientists listen allows us to challenge what science is, where science happens, and who can be considered a scientist—Katy Payne, after all, never received formal training in biological sciences, yet her work would spur on the burgeoning discipline of bioacoustics. Indeed, this study of more intuitive ways of knowing in fieldwork brings into relief how what is published as "science" or who can be considered a "scientist" has excluded women, people of color, and nontraditional ways of knowing for centuries.[21] This erasure persists, in part, because studies of science tend to focus almost exclusively on texts, or those scientific findings that have been published and distributed.[22] By contrast, highlighting how what is rhetorical and significant about science far exceeds academic texts or a study's conclusive findings, allows for more inclusive parameters of what constitutes scientific work and knowledge. Rather than superfluous anecdotes, this study considers all the imaginative, creative process of ideation and attention that goes into scientific studies as intellectual resources that allow researchers to listen and know otherwise.

Listening, then, is a quotidian element of rhetorical invention in the sciences that remains an implicit, under-theorized feature of scientific method. Following communication and media theorist John Durham Peters, who differentiates between techniques and technologies, I understand scientific method as deeply theorized in its technologies and under-theorized in its techniques. Peters contrasts techniques from technologies according to the durability of a mode's materials. For example, writing—words on stone, page, screen, which can endure for centuries—is a technology, whereas speech, fleeting if not recorded, admits technique.[23] By this definition, scientific method is filled with technologies that insist on recording and documenting its methods for the sake of replicability in future experiments. Observation and listening are techniques used in the technological process that is scientific method, and scientific method always requires an interplay of these techniques to promote multisensory practices.[24] In other words, an interplay of the visual, the aural, the tactile, the gustatorial, and the olfactory converges to extend, expand, and complicate ways of knowing during any scientific study.

Science's advances are often studied through the progress of its technologies, but the researchers in this book exemplify how science's primary instrument remains the body of the scientist—the stories of how their bodies are changed over the course of their studies are stories about techniques for listening. Because *Listening to Beauty* is interested in how biologists can sense through the chest, the ears, and the feet, I reconsider the traditional scientific observer through a notion of listening that will expand with each passing chapter through various definitions and redefinitions of the sensory mode most associated with the ear. By the end of this project, we will come to understand listening as an umbrella term that encompasses the bodily techniques of scientists in the field—listening will have as much to do with hearing as with seeing and feeling.

Yet those messages received through listening may require translation. When it comes to scientific research, translation must come through words and symbols in order to be conveyed to others. These moments of translation mark when listeners become witnesses. If witnessing is about giving an account of those messages received from nonhuman kin and our natural environments, then the scientist as "witness" accounts for the interplay of sensory modes and imaginative, intuitive ways of knowing during rhetorical invention. Witnessing helps us get at the sensory exchanges involved in making meaning on "both sides" of language, the symbolizing and the receiving, the discoursing and the listening. To put it simply: witnessing allows relationality at every turn. But unless we fully define the role of listening in witnessing, we risk falling into the harms of witnessing that feminist science studies scholar Donna Haraway articulated long ago. The scientist as "witness" here need not be a modest witness, but may be an empathetic witness, a witness caught up in the messy entanglement of being there with their subjects of study.

Witnessing

Scientific method has always required witnesses, or those who can confirm that proper procedure was followed, that results were reported reputably, that the facts established by certain studies have earned credibility. Historically, for the weight of such testimony, not just any witness would do. The beginning of institutional, scientific writing ushered in the emergence of the ideal witness as the "modest witness," or someone who removes their feelings, emotions, and subjective interpretations from consideration and who relays their findings in language that is straightforward, descriptive, and technical.[25] This idea that a modest witness must render themselves invisible and deem objects of study free from influence in order to establish objectivity is one long challenged by feminist science studies.

Theorists like Donna Haraway, who famously termed objectivity as a "view from nowhere," have argued that because scientists can never escape their situated standpoints, the best and strongest scientific information is an account that acknowledges the power and positionality of those scientists.[26] The scientist as modest witness is one "endowed with the remarkable power

INTRODUCTION

to establish facts," one who "guarantees the clarity and purity of objects," Haraway outlines, somewhat sarcastically.[27] By contrast, the scientist who is not afraid to admit the limits of knowing, does not pretend that such limits can be overcome by disavowing their bodies and biases, is a witness still, but a "feminist, antiracist, mutated" witness queered through a conception of objectivity that accounts for where scientists see from.[28] The researchers in this book bring into relief an even fuller conception of witnessing in the sciences, one that accounts for how scientists listen and how sensory, intuitive knowledge serves as an epistemic resource in scientific studies.

To put it another way, more than a mere technique of observation, science's "modest witnessing" is an aesthetic regime that guides how scientists sense their subjects of study. Any historical notion of aesthetics, argues literary scholar and critical race theorist Kandice Chuh, centers on the idea of "the sensus communis," or a way of sensing and judging in common.[29] What is considered political in rhetoric, objective in science, and beautiful in art is determined by how others in a shared community or discipline would judge what is political, objective, and beautiful. For Chuh, the fact that aesthetics stemmed from a narrow sense of shared taste or judgment ensured a racist, gendered, classist regime in which some bodies are valued as more credible arbiters or witnesses than others; as she puts it, the inability to make a "proper aesthetic judgment signaled the difference between those who would and would not realize human potential by achieving full self-consciousness."[30] Because proper aesthetic judgment in the sciences remains too closely tied up in the modest witness, *Listening to Beauty* consults researchers who employ intuitive, nonrational modes of sensing in the field and who thus have much to teach about broader, more empathetic methods of witnessing.

Further, because the biologists in this book empathize with the experiences of nonhuman animals, this study requires an approach to witnessing that decenters the modest human witness and values the aesthesis of nonhuman beings. For rhetoric scholar Casey Boyle, "reconceiv[ing] witnessing beyond the humanistic orientation" opens the aesthetics of witnessing not "in terms of mono-perspectival and mono-temporal accounts, but rather [through] multi-modal and multi-temporal operations."[31] Important for our study is that even as we retain the centrality of witnessing in the sciences, we contest that scientists are the only witnesses in their studies and thus not the only individuals who may give testimony to what occurs in natural environments. No coincidence, then, that as I queried each researcher in this book about their mode of sensing and knowing the testimonies of their nonhuman kin, each turned to "beauty." More than a marker of proper aesthetic judgment, beauty breaks open dominant aesthetic regimes—in the cases to come, beauty gives scientists permission to refigure their field of experience, opening scientists to different forms of analytical possibility. Beauty attunes scientists to the shared arena of sensation in which they may meet with their nonhuman kin. Beauty opens a crack in perception, issues a call to sense otherwise—an imperative to witness.

This articulation of scientists as witnesses leverages a different conception of witnessing, one diffracted as much through Haraway and feminist science studies as through theories of witnessing in rhetoric, which more explicitly articulate the role of aesthetics in giving accounts of memory and testimony. Such reports of witnessing help to decenter the role of visual observation in scientific method to make room for listening—no longer are scientists exclusively observers; instead, scientists are encouraged to get caught up in the multimodal milieu of invention in which sensory modes—not just seeing and listening, but also touching, feeling, smelling—combine to place the scientist in the throes of aesthesis. Yet giving an account of experience is no simple task; Peters even goes as far to say that witnessing is "an intensification of the problem of communication more generally" in that witnessing presupposes a gap in knowing, a discrepancy in understanding between two parties for which one must give account, an epistemic hole that requires filling.[32] Because we fill epistemic holes with words that attempt to narrate our experiences but can never narrate the whole of those experiences entirely, how we witness, or make present past experience, is, by nature, faulty.

Words help us to define, pinpoint, and refine meaning, but excess of sense, feeling, emotion, and affect inevitably spills over. Then again, meaning itself is a rhetorical problem in that our tendency to assign meaning only to that which has been defined through words limits our ability to value nonlinguistic meaning that comes, already fully articulated, through song, dance, and other art forms, not to mention intuitive and imaginative ways of knowing. The "mistaken idea that meaning is solely, or even mostly, a linguistic phenomenon" is one that should be abandoned, according to philosopher Mark Johnson.[33] Johnson extends meaning aesthetically, defining cognition as embodied and emotional, embedded and extended into environments, evolutionary and active, shifting, and larger than any individuals' discrete modes of processing. Applying these problems of defining meaning and valuing aesthetics to the idea of the "modest witness," we could say that if the sciences are to value more than discursive meaning and nontraditional ways of knowing, then even objective, data-centric fields must reconsider their relationship with aesthetics.

Put simply, witnessing need not be so modest, and some techniques of witnessing are more aesthetic, or sense-inflected, than others. The aim of

Listening to Beauty is to show the affordances of different methods of witnessing, to uncover how sometimes it takes a more empathetic witness to bring forth scientific knowledge. Take the case of geneticist Barbara McClintock, who, as physicist and philosopher Evelyn Fox Keller reports, patiently studied individual plants, kernels, and chromosomes until she had "a feeling for the organism."[34] This feeling arrived intuitively, wordlessly for McClintock, who eventually convinced others and won a Nobel Prize for what she felt and knew to be true: that genetic elements can change position on a chromosome and so are more dynamic than previously understood. As McClintock put it, "You work with so-called scientific methods to put [what you feel] into their frame *after* you know." All the while the question remains, even for McClintock, "*how* you know it."[35] How scientists get a feeling for organisms, how scientists study beautiful subjects in the field, and the *how* of science in general are all questions about witnessing, about making meaning—in other words, about aesthetics.[36] Aesthetics is the arena where words and affect meet, where we study the interplay between what can be articulated and what can only be felt, where we consider all that is left over when we explain something away. Here—at the expanse of aesthesis, at the edges of discourse, in the throes of listening—science is at its most rhetorical.[37]

Field Stories

The more listening came up in my interviews with these biologists, the more attention I paid to my own methods for listening, not only in our conversations, but to all the moments, scenes, and interactions surrounding my time with them. If I was asking scientists to be honest about their approach to their subjects of study, then I, too, should be honest about my approach.[38] In *Listening to Beauty*, I practice what anthropologist Sarah Pink has called "sensory ethnography," which is a way of being there with what, who, and how we study. Sensory ethnography asks researchers to consider qualitative interviews as "social, sensorial, and affective encounters" that require being attuned to the sounds, tastes, images, and bodily gestures exchanged during the "conversation" as much as the actual "talk" of the interview; Pink advises that these ethnographies should be presented to readers through rich, sensory description.[39] Through sensory ethnography, I allow the theory that emerges from this project to be grounded in lived experience, not only through how biologists describe feeling their way through field environments but also how my calling attention to sensation, feeling, and aesthetics in their work unearthed new understandings of how they rhetorically encounter nonhuman kin and their natural environments.[40] Narrative abounds in this text to account for what happens at the limits of language—these tales chronicle how certain biologists learned to listen.[41]

In the chapters to come, biologists' accounts intermingle with details on nonhuman ways of knowing—how whale songs rhyme, how elephants communicate through their feet, how birds make song through their wings. You may think, as these stories unfold, that I am anthropomorphizing, ascribing human thoughts and emotions to these animals, and maybe I am. Or—and imagine with me here—whales bear witness, birds remember, elephants' bodies, like our bodies, hold generational memory. Rhetoric scholars have long abandoned the idea that verbalizing is the only way to give an account or offer testimony, so perhaps we should not consider our human memory as the best memory. Taking beauty seriously thus allows us to consider not only listening and witnessing, but also notions of anthropomorphism and objectivity in a new key.

Chapter 1, "Nature's Punctum," starts there—with the rhetoric and philosophy of aesthetics in a new key. As has been established, rhetoric in this study is no longer bound to a human rhetor but is distributed to landscapes, animals, music, matter, and other nonhuman agents. This chapter lays the theoretical groundwork for such a project by locating a different historical touchstone for new materialist approaches to rhetoric and science studies in the work of Susanne K. Langer. A contemporary of Thomas Kuhn and a student of Alfred North Whitehead, Langer devoted her life's work to understanding how the human mind evolved to interpret symbols, maintaining that scientists, like artists, must begin their work through the resources of feeling. Relying on Langer to extend Roland Barthes's famous reference to the "punctum" as an affective element that reaches out to grab viewers' attention, this chapter presents "nature's punctum" as that beauty which reaches forth from natural environments to puncture scientists' logic and bring them to feeling. This chapter thus shows how what is rhetorical about the sciences—that which persuades human scientists to pay attention—exceeds, and possibly precedes, the bounds of discourse and disciplinary ways of knowing.

Chapter 2, "Punctive Listening," grounds this theory of nature's punctum through an intensive interview with Katy Payne about her breakthrough revelation that humpback whales were singing. The understudied story of the songs of the humpback whale is that Payne, a music major in college, has never considered herself a scientist. This chapter examines how, rather than pursuing hypothesis-driven research, Payne employs contemplative practices that promote deep listening to her surroundings and the beings therein. As she listens to a semiosis other than her own, Payne exemplifies a way of witnessing that is more felt and empathetic than modest.

Yet because bodies best learn how to sense natural environments with which they are familiar, we must learn the limits of listening in scientific practice. In that vein, chapter 3, "Extractive Listening," takes up Payne's legacy in contemporary bioacoustic science through interviews with the researchers who make up the Elephant Listening Project at Cornell University and who detail what the ear of Western scientists may miss while studying elephants in African rain forests. Named after Payne's second revelation that elephants communicate via infrasonic signals, the Elephant Listening Project's lead scientists Peter Wrege and Daniela Hedwig employ acoustic monitoring to decode elephant rumbles in African rain forests and determine whether elephant communities have developed methods for warning one another about poachers. Increasingly more grounded in quantitative scientific practices, Wrege and Hedwig grapple with the humble realities of experiencing the beauty of a species on the verge of extinction. Often depicted as the most emotional nonhuman animal, elephants provide a unique opportunity for studying how dominant scientific practices risk listening hungrily and extractively if they are not listening with and alongside local communities and nonhuman kin.

Chapter 4, "Emergent Listening," returns to the ocean, the medium in which sound travels fastest, lingers longest, and has the most to teach us about the pain of listening. Here, we learn from marine biologists Laura J. May-Collado and Michelle Fournet, who present the "anthropause" of the COVID-19 pandemic as proof of the degree to which human noise stresses marine mammals who are unable to escape boat motors, offshore drilling, and seismic surveys. Considering the global impacts of how heavily humans tread, this chapter analyzes the etymological roots of "emergency" to uncover what it means to sound sirens and to listen in the urgent time of climate change.

By contrast, chapter 5, "Swallowed by Beauty," finds beauty in noise—noise in natural environments, that is—as biologist Patricia Brennan details the "rowdy" environment of Amazonian forests as one that grants scientists the feeling of "being swallowed" by a force greater than themselves. If scientists are to challenge disciplinary ways of knowing, "being swallowed" by beauty, by broader ways of listening, may be a necessary first step. Yet the book closes by considering the risk and controversy currently surrounding beauty in evolutionary biology through ornithologist Richard O. Prum's attempt to amend Darwin's theory of evolution by centering it on aesthetics. Should evolutionary biologists employ feminist ways of knowing, Prum argues, then they would have to consider female birds as agential subjects whose preferences are worth attending to. Accounting for female subjectivity brings Prum to argue that beauty is not a measure of fitness but a key insight into the lived experiences of our nonhuman kin.

By way of conclusion, I consider whether the argument about listening to beauty is bound only to mammals we already know and love. Entomologist Ron Hoy—who has spent the latter part of his forty-year career studying the *Habronattus* jumping spider, which has been called "the bird of paradise of spiders"—locates the same lessons about beauty in insects. Born into an immigrant family, Hoy further details a life of "seeing from the margins" and articulates what Payne, elephant rumbles, and the noise of the Amazon make clear: science is at its most beautiful when it is at the margins of itself.

For too many of these researchers, this talk about beauty puts their reputation on the line, makes them the punch line of old jokes about rigor and traditional notions of objectivity. One biologist cannot speak for the rest, but when I asked Prum why he risked talking about beauty anyway, he returned with a different question: "Where has science gotten us?" Dominant science has brought electricity, vaccines, and sustainable energy, but also microplastics, eugenics, and the greenhouse gases accelerating climate change. My proposal is that a turn toward the beautiful is not a turn away from the technical. Instead, a turn toward the beautiful—to the intuitive, imaginative, and the aesthetics of invention—may just offer a better, more careful turn to the technical. For those of us not squarely situated in the sciences, may we learn how the resources already in us, the bodily practices we employ to understand our environments, prepare us to listen, to witness, to ask the kinds of questions that may just challenge scientific ways of knowing. May we learn to value all the expertise we bring with us when we enter these vast worlds.

1

Nature's Punctum

> The most productive thinkers of those maverick days [of science] ventured on some wild flights of fancy. Not so the founders of the "young sciences" today. They cannot indulge in fantastic hypotheses about the aims or origins of society, the presence of sentience or intellect in anything but the investigator himself.
>
> SUSANNE K. LANGER, *Mind: An Essay on Human Feeling*

> Now halfway through [Langer's *Philosophy in a New Key*], nearly at the end of the section on Language. (I confess that I have never read this book, although I should have. I have a copy that I began quote some years ago—but as is often the case, I got sidetracked, and never returned to it until now.)
>
> KENNETH BURKE, note tucked inside *Philosophy in a New Key*

Just outside of Andover, New Jersey, an aged, wood-shingled house sits, pretty much as it was left in the early 1990s. Red-painted trim chips away from the kitchen window that overlooks the lawn where William Carlos Williams, Malcolm Cowley, and Ralph Ellison once gathered. The musty smell of old books overwhelms those who climb the wooden staircase inside the house, which was once the home of Kenneth Burke; floor-to-ceiling bookshelves cram the space, lining the upstairs hallway, the walls of every bedroom, and even some closets. Overflowing with books—leather-bound from the 1920s and '30s, paper-bound from the '40s onward—these shelves are full of ideas written at the hand of perhaps the most well-known rhetorician of the modern Western era. To this day, Burke's books are organized exactly as he left them. And if you look closely, you find that a select few books have pages of Burke's notes, folded lengthwise, sticking from their tops. That's how, when I started paying attention to those books Burke left overflowing with notes, I pulled the work of Susanne K. Langer from the shelves.

Burke argued that almost everything—art, literature, myth, science—was rhetorical, persuasive, and capable of shifting logics, and Langer, his contemporary who studied philosophy under the tutelage of Alfred North Whitehead, crafted similar messages about the importance of rhetoric in the postmodern era.[1] Yet whereas Whitehead and Burke remain key figures in the history of philosophy and rhetoric, the same was never the case for Langer, probably because it was too easy for readers like Burke to get "side-tracked" while reading lesser-known women philosophers. At least Burke realized, while writing his infamous "Definition of Man" essay, that he should have read Langer's work sooner.[2] This chapter imagines what may have happened to theories of rhetoric, aesthetics, and science studies had these fields also read Langer—and taken her seriously—sooner. This chapter will set Langer into relief as a thinker who teases out the relationship between the sciences and aesthetics, or names how it is that aesthetic processes allow us to make meaning alongside our natural environments. Langer draws our attention to how the rhetorical work of engaging one's environment is not always so rational; through her writing, we come to understand why when the biologists in this study face the limits of discourse and disciplinary logic, they turn to beauty.

Trained as a mathematician and logician, Langer would maintain a creative tension with the sciences throughout her career. For example, she received invitations from the Committee on Space Efforts and Society of the American Academy of Arts and Sciences to weigh in on how their program might "avoid damage to society"; she was approached by biologist C. H. Waddington to join a league of thinkers—biologists, physicists, mathematicians—at an international interdisciplinary symposium on the theoretical foundations of biology; she fielded manuscripts to review from cognitive psychologists like Howard Gardner, whose theory of multiple intelligences would go on to alter how psychologists and educators consider individuals' approach to learning and knowledge creation.[3] Another time, Langer received a distinguished scholar invitation from the American Academy of Arts and Sciences to deliver a response to papers presented at the Conference on Science and the Modern World View, and her summation of the conference was that the arts and the sciences were much too divided:

> The person who is steeped in the fascinating work of science is a layman in aesthetics.... Similarly, the person who is immersed in a study of English ... has only a popular conception of what biologists are doing, and a vague suspicion that they will reduce the concept of Life to some mechanical concept.[4]

For Langer, arts and sciences remain limited when each discipline is held in opposition to the other. Because she prioritized interdisciplinary thinking, she was one of the first thinkers György Kepes approached when he began building the arts program for engineers at the Massachusetts Institute of Technology. Through these engagements, Langer offered a theory of symbolism grounded in and exigent to the biological sciences.[5] From Langer, we begin to understand what happens when the evolutionary biologists profiled in the following chapters encounter and learn from nonhuman ways of knowing. In the chapters to come, we will encounter the sounds and signals of whales, dolphins, elephants, birds, even jumping spiders, but first we must set a foundation through which to understand those encounters as moments that appeal more to the skin, the gut, the ear, rather than the logical corner of the mind. These more aesthetic appeals, Langer taught, are as articulate as discourse.[6]

Moreover, Langer was convinced that if biologists—and scientists more generally—returned to their more aesthetic foundations, then they might better understand how human scientists evolved to feel and sense in the first place and engage with natural phenomena anew. As she insisted: "the most productive thinkers" in the history of science were those who "ventured on some wild flights of fancy."[7] What happens during those "wild flights of fancy" I will come to define as an aesthetic, punctive encounter in which the resources of an illegible intelligibility come to the fore. In *Camera Lucida*, Roland Barthes famously defined "the punctum" as an element in photographs that puncture, or sharply resonate with human viewers, in a more than rational way. The punctum conveys meaning through an affective intensity that can never fully be articulated. Concerning the ineffability of the punctum, Barthes remarked that "what I can name cannot really prick me."[8] I use the idea of the punctum, then, to complement Langer's philosophical contributions to the sciences. Thinking with Langer, I theorize what happens affectively and symbolically during creative moments of invention, and extending these ideas through Barthes's punctum, I articulate *how* the biologists in this study can be imposed upon, how their logical systems may be disrupted, and how they might come to know otherwise. To put it another way, the punctum persists beyond artwork, and what I will define as "nature's punctum" plays an integral role in how we sense, know, and make kin in our natural environments. If the rhetorical turn to "beauty" is an attempt to make sense of, convey, or attend to those punctive encounters, we must first understand what happens symbolically and bodily—not just visually and logically—when biologists set out to study "life" in the field.

Diffracting Disciplines

Because her academic training crossed and challenged disciplinary lines, Langer was in an ideal position to detect where the sciences reached their limits in attempts to explain human feeling. In the postwar era in which she began lecturing in philosophy departments, Langer assessed that positivistic science, which rejected intuitive or metaphysical approaches to knowledge, had come to inundate most, if not all, intellectual endeavors, even humanistic ones. Her work, then, was invested in articulating a model of feeling that outlined the resources of imagination, that encouraged and guided the intuitive hunches at the start of any intellectual endeavor. Her ambidexterity of thought, however, contributed to Langer's difficulty in gaining notoriety during her own and the present time.[9] One reason that a thinker like Langer falls through the canonical cracks or stays stuck on dusty old bookshelves is that she remained in the shadows of Whitehead, her famous mentor. Another reason is that Langer was one of the first women in the United States to live out an academic career in philosophy. But the reason that eclipsed her teacher and the gendered norms of her time was that Langer, an expert in weaving together different lines of thought, left her critics—comfortable in their disciplinarity—confused as to what she was up to. Today, we might consider Langer a key thinker in the transdisciplinary subfields of philosophy, history, and rhetoric of science, but because her writing appeared decades before Kuhn's *The Structure of Scientific Revolutions*—the 1962 landmark publication that called scientists to attend to their social, cultural, and rhetorical commitments—would highlight the importance of those subfields, her work seemed hard to place in her era. Although she was most firmly committed to the discipline of philosophy, Langer exemplified what feminist philosopher of science Iris van der Tuin calls "diffractive thinking."

Diffraction is a pattern that emerges as a result of interference, witnessed when a stone is dropped into a pond or when the wakes of two boats overlap. As waves overlap and interfere with one another, they come to cohere, creating a pattern of resonance from within a pattern of dissonance. Diffraction—as a methodology—allows distinct perspectives, concepts, or philosophies to intra-act, with the aim of reflecting the points of merger as well as the points of divergence between those disparate perspectives.[10] To think diffractively, then, is to disrupt hegemonic thought by twisting established, canonical lineages amongst and betwixt one another. By refusing to tether her concepts to any traditional lines of philosophical inquiry (e.g., tracking a concept such as aesthetics through Hegel, Kant, Husserl, and others), Langer freed herself to pursue nontraditional lines of inquiry from many disparate schools of thought like anthropology, art history, evolutionary biology, music, and psychology, to theorize aesthetic experience.[11] For Langer, intellectual history was not hierarchical or chronological, an unfolding of ideas that tracked from one intellectual position to the next. Instead, ideas folded back on themselves, with aesthetic concepts like "feeling" and "vitality" diffracted through scientific ways of knowing and resurfacing with new material meaning.

For example, in *Feeling and Form*, Langer theorized how vitality and affect (feeling) are held by artwork (form); in doing so, she problematized the dichotomous association of feeling with spontaneity and form with controlled

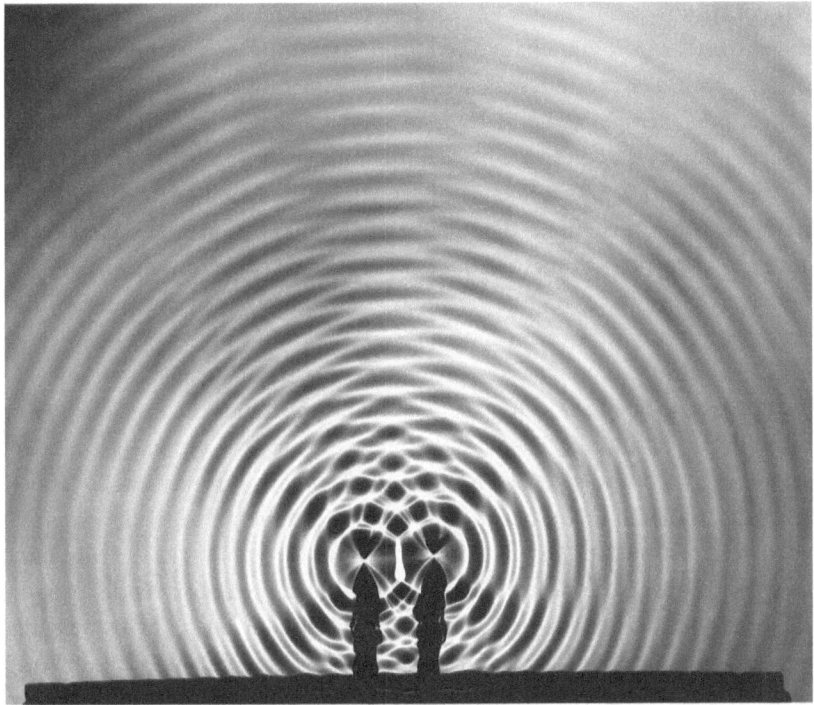

FIGURE 1. Diffraction patterns. In Karen Barad, "Diffractions: Differences, Contingencies, and Entanglements That Matter," in *Meeting the Universe Halfway: Quantum Physics and the Entanglement of Matter and Meaning* (Duke University Press, 2007), 77. Used with permission from Karen Barad. Photo by Berenice Abbott/Getty Images.

regulation.[12] Through rich analyses of sculpture and music, Langer considered our human, subjective experiences of artwork—that most common concern in aesthetics—but was more interested in the role of objectivity in art, how artwork carries and conveys feeling. Art is the objectification of feeling, she argued, and which feelings come to be objectified stem from our evolutionary history. In other words, artwork can evoke and bring us to know—intuitively, if not discursively—a deeper, more fundamental symbolism within us. This symbolism is more than discursive, an affective symbolism that explains how beauty compels attention.[13] Indeed, less concerned with highbrow conceptions of art or beauty as indicative of taste, Langer, like our present line of inquiry, was concerned more with how it is that beauty convinces us to maintain our focus.

This affective symbolism or this illegible intelligibility—our ability to know aesthetically through the senses—is experienced not only in artwork, but also in natural environments, precisely because art is not separate from, but constitutive with, nature. According to Langer's evolutionary theory of the human mind, if humans evolved from that nature we seek to understand, then any human creative endeavors, such as art, also derive from nature. Langer thus exemplifies a post-Kantian approach to aesthetics in which experiencing beauty is a conduit to certain methods or approaches to praxis rather than an ideal end, an experience of proper, aesthetic judgment. That is, for Kant, the beauty experienced in nature was different from the beauty experienced in artwork, with nature's beauty as a more pure, primary aesthetic experience.[14] For Langer, nature's beauty is not separate from beauty in artwork. As she articulates, "the connection with the natural world is so close" that art is the "subjectification of nature," or one way nature makes itself known.[15]

In *Philosophy in a New Key*, Langer would speak this message directly to those in the positivistic sciences: this new key, or new approach to understanding symbols, was needed because the "heyday of science . . . has stifled and killed our rather worn-out philosophical interests."[16] Yet empirical science need not be the end of imaginative, intuitive inquiry, as the substance of empiricism derives from the same evolutionary semiosis as philosophy, rhetoric, and art. Put simply, humans rely on discourse to convey meaning, but discourse is not the only way to convey knowledge and often fails to bring into relief all the inventive, imaginative, affective excess involved in human experience. Langer then theorized an expansion to the range of symbolism. In addition to discursive symbols—texts, speech, and other verbal forms—Langer identified "presentational" symbols that convey meaning through image, sculpture, music, and other art forms. "Non-discursive symbolism" is another name Langer used to explain the work of presentational symbols. Opening symbols to exceed the discursive allows us to make sense of how affect, intuition, and other ways of knowing can persuade, how we humans are moved by sensation, feeling, and so much that we are not able to verbalize. Nevertheless, because terms like "non-discursive" define the richness of feeling and intuition by what it is not—discourse—I will amend Langer's phrasing in what follows to consider the discursive and the "more than discursive," or the discursive symbol and the affective symbol.

Saying that the biologists in this study rely on more than discursive symbolism in their work is another way of insisting that aesthetics must be brought to bear in order to understand rhetorics of science. Because, as Langer put it, "what is directly observable is only a sign of the 'physical fact'; it requires interpretation to yield scientific propositions." And here, she does not just mean textual, verbal interpretation: "not simply seeing is believing, but *seeing and calculating, seeing and translating*" is what makes science, according to Langer.[17] What is rhetorical about science broadens even more when we take seriously how symbols in all their subjective and affective messiness are caught up in scientific methods of inquiry. "The problem of observation is all but eclipsed by the problem of *meaning*," Langer posits, "and the triumph of empiricism in science is jeopardized by the surprising truth that *our sense-data are primarily symbols*."[18] To be clear, the symbolic meaning of sense-data is not conveyed discursively; rather, it is communicated subconsciously, bodily, affectively. Langer's symbolism thus challenges scientists to account for the multiple symbolic layers in their studies, the most fundamental of which may be a sort of kinship between researchers and their subjects of study.

Quite simply, Langer's theories allow us a foray into those early moments of thought in which scientists come to know their subjects of study intuitively. Her symbol system was thus less about one-to-one representation (i.e., that preoccupation of most semiotic theories) and more about the nonrepresentational and material values that inform and inflect our ways of knowing.[19] Her diffractive exploration of symbols exposed how humans come to feel and make sense of the evolutionary lineages within and around them. In her own way, Langer, too, insisted that such lessons may best be learned through listening.

The Sound of Prescientific Invention

You can always close your eyes, but it is not as easy to close your ears—sound seeps in before we can avoid it, catches us unaware, moves us to feeling. This is the reason that music was the best teacher for Langer; music proves her point that some forms, some symbols, need no linguistic interpretation.[20] When we give ourselves over to listening, Langer expected that we will realize how "music articulates forms which language cannot set forth."[21] And it is not that translation is needed: "art," she explained, "certainly music, and probably all art—is formally and essentially untranslatable" through discourse. "It is not true," she further argued, "that whatever can be expressed symbolically can be better expressed literally. For there *is* no literal expression, but only another kind of symbol."[22] Music thus exemplifies that affective symbol that presents information through immediacy of feeling.

No surprise, then, that many famous scientists were first musicians. Einstein, gifted at playing the violin, famously stated that if he had not been a physicist, he would have become a violinist.[23] Geneticist Barbara McClintock played the tenor banjo in a jazz band during her undergraduate studies.[24] More recently, evolutionary biologist Cassandra Extavour made headlines in *Nature* for her side career as a soprano.[25] Langer lived in her own world of music, playing piano from an early age and studying cello and music composition in college.[26] What Langer experienced with music composition—how music moves us to thought through feeling—she also considered as essential to how scientists come into knowledge. Langer would name this process of idea formation as the "prescientific stage" in scientific inquiry, a stage that rhetorical theorists would call "invention." Recall that rhetorical invention references the gathering of ideas to prepare an argument or find the available means of persuasion. The trouble with articulating what, exactly, happens during this stage of rhetorical invention is that processes like idea formation are difficult to detail in words; invention involves a sensory, affectively laden process that is as visual, haptic, gustatorial, and sonic, as it is verbal. In other words, when Langer queried the imaginative processes that occur "before" conceptual scientific inquiry, she was querying what we may consider "prescientific invention," or those early moments when science is led by feeling and intuition.

Prescientific invention articulates the affective milieu that surrounds and enlivens discursive, disciplinary thought.[27] The problem for Langer was that scientists all too often skip this stage of invention or shelve the feeling in order to push toward discursive interpretations of data. Usually, scientists are preoccupied with what Langer called "idols of the laboratory." If Francis Bacon's "idols of the mind" explain how adhering too closely to cultural scripts can cloud our perception, then Langer's "idols of the laboratory" explain how scientists err when they grant too much weight to their terms and, moreover, extend mechanistic ideals to explanations of the emotive, affective elements within human encounters, not to mention explanations of aesthetics in natural environments.[28] Idols of the laboratory are not unlike what scientists today call "physics envy," which is the tendency for biologists, social scientists, and even humanists to make their research out to be as generalizable and rigorous as highly mathematical fields like physics. Further, relying too much on disciplinary knowledge that has come before risks never encountering our subjects of study in their fullness, never being open to nature's surprising revelations.

Should the sciences return to feeling, Langer implied, then they may more fully encounter and understand nature's forms—of which the human mind is a part—because during this prescientific stage of heightened imagination, scientists come to feel the vivid values that emanate from their environments. Here, Langer echoed a key tenet from Whitehead, who hypothesized that prior to major shifts in knowledge, there occurred "an intense period of imaginative design" required to break through the brain's habit for rote processing.[29]

Whitehead offered the example of a sunset to explain what happens during prescientific invention, or these moments of heightened imagination:

> When you understand all about the sun and all about the atmosphere and all about the rotation of the earth, you may still miss the radiance of the sunset. There is no substitute for the direct perception of the concrete achievement of a thing in its actuality. We want concrete fact with a high light thrown on what is relevant to its preciousness.[30]

To argue that scientists navigate their environments with a sense of how "precious" it is does not highlight sentimentality; instead, "preciousness" carries the idea of "non-material value."[31] Which is another way of saying that scientists should aim to articulate facts grounded in concrete matter, as they are known to do, but they should also account for matter's "vivid value," or how it is that matter conveys its vital energy to us.

Langer's broad definition of the symbol aids in this process of negotiating how aesthesis emerges from matter, how feeling emerges from nature. Inherent to her semiosis is a certain conception of how, exactly, humans interact with and sense the materiality of their surroundings. We like to pin down words and symbols, especially in the sciences, to achieve shared meaning, but for Langer, the indeterminacy of words is an inventive resource. Because symbols serve as the bridge between the ideal and the material, there is constant negotiation between the individualized, sensory, felt experiences of words, and the environmental material that compels and inspires our words. Consider how science happens long before its contributions can be articulated, and even then, how scientific exactitude is often at odds with language, as Langer argues:

> The most searching linguistic studies give us no elements comparable to the things scientists talk about—atoms, molecules, or measures of energy in other forms. Words are incorrigible weasels; meanings of words cannot be held to paper with the ink. The abstraction of pure concepts may occur under stringent controls of technical terms, but the inveterate tendency of even such terms to become assimilated to common parlance and share the ways of "ordinary language" leads one to wonder whether the great frame of science can possibly be made of word-borne thought. The formulation of "fact" may stem from language, but "science" in the modern sense does not.[32]

Scientific "facts" attempt at shared objectivity, and Langer would likely agree that such shared symbolicity is both possible and necessary, but so much felt affect—that which is known in more than rational ways—exists beyond that traditional notion of objectivity. These aesthetic elements of the sciences that cannot be pinpointed by "incorrigible weasels" like words must be felt, perhaps primarily through listening.

In moments of prescientific invention, the verbal, the visual, and sensory processes more generally intermix in a swell of aesthesis that may best be known intuitively through sound. Consider how Black feminist theorist Tina Campt calls to visual scholars to "listen to images." Sound registers across the sensorium and need not be heard to be felt, Campt explains, which means that sound is sometimes, but not always, sonic.[33] Sometimes sound is vibrational or even imagistic; even when sound is heard, it consists of more than what is heard. Lower frequencies, like those often present in animal communication, are haptic, registering more immediately as a hum, or a rumble, than as "sound." Campt finds these low frequencies in images, an occurrence she describes as "felt sound."[34] Placing Campt in conversation with Langer, we can say that felt sound—information that resonates bodily, affectively, intuitively—may be the primary register of prescientific invention. Indeed, as Langer's writing about music suggests, our eyes may be the worst guide when it comes to feeling.[35]

Langer's theory of prescientific invention allows us to return to rhetorical invention and define it as a more than discursive experience that may give way to discursive logic but that may also open us to asignifying ruptures that suspend us in feeling and leave us to withhold interpretation. Withholding interpretation, listening to sound, and listening, also, to sound that can only be felt become ways of listening that are not listening *for* something, but listening to what we do not, and possibly cannot, know.

Feeling Nature's Punctum

That humans are foremost feeling beings is perhaps the central thesis of Langer's oeuvre. As American art critic Arthur C. Danto describes Langer's overarching argument: "Because we are never not embodied, feeling is our essence as human beings, and rational thought but one of its more perspicuous modes."[36] The capaciousness of feeling for Langer has everything to do with how sense information is "primarily symbols." The symbolism of the senses zeros in on how the immaterial is wedded to the material for Langer: feeling is what connects sensation to abstract thought or, in the case of art, what grounds abstract thought into a material, moving body. Feeling, then, is registered through symbols—not discursive symbols but more than discursive symbols—whose meaning punctures.

To be known discursively, that which punctures must be diffracted through other communicative modes.[37] Langer would come to term this ability of art to convey feeling and life to viewers affectively as "felt life." And because felt life punctures through feeling, it is an action—not a reactionary, passive mechanism, as it is often depicted. Langer also never sets a dualistic opposite, such as "unfelt life" versus "felt life." For her, all life, even the evolutionary life history within us, may be accessed and felt, even if not in discursive modes. She insists that art, writ large, brings even "the felt tensions of life, from the diffused somatic tonus of vital sense to the highest intensities of mental and emotional experience" to the forefront of attention and evaluation. This "tonus of vital sense" relates to the lowest level of muscle energy that is always somatically firing—even that "tonus of vital sense" may be excited and felt through music, art, and nature.

Notions like "felt life" were likely what brought scientists, more so than artists, to offer the "most sympathetic reading" of Langer's work.[38] Biologists were particularly attracted to Langer's theory of feeling because she conceived of feeling through how humans evolved to experience life. Her evolutionary theory of aesthesis in *Mind* even compelled one biologist to describe her work as "the kind of careful and reflective observation which the scientist admires."[39] Langer considered her work the same way. When another reviewer of *Mind* referenced her work as "a metaphysical system," Langer remarked that the reader should "know the difference between a 'metaphysical system' and a scientific line of inquiry."[40] Discussing abstract thought for Langer was not grappling with metaphysics; instead, thought was part of that life that could be "felt."[41] "Felt life" helped Langer to explain how aesthesis bubbles forth to the surface of feeling.[42] This aesthesis was deeply biological, material, and cerebral—a thinking inflected with aesthetics, a working out of sense.[43]

"Felt life" and its earlier iteration as "vivid values" remind one of Barthes's "punctum" in how these terms portray a moment of sharp puncturing that resonates with us—typically as human viewers—in a more than rational way. In *Camera Lucida*, Barthes proposed that two elements exist within any photograph: the studium and the punctum. The studium is the mix of cultural symbols through which the photograph accrues and becomes coded with meaning. The punctum, on the other hand, is not so discursive.[44] It works more affectively, reaching out to grab viewers as the "element which rises from the scene, shoots out of it like an arrow, and pierces me."[45] Together, the "studium" and the "punctum" have come to be considered as a semiotics of the visual that "challenges the very nature of the notion of 'rational' by taking the corporeally lived, emotional aspect of the experience as an integral component of signification."[46] In other words, even if photographs may not always mean conceptually, they mean according to how they bring our bodies (or bodyminds) to feeling. The punctum, specifically, strikes viewers in such a way because it means, or affects, through an affective intensity that cannot be articulated fully in words.

Articulate meaning and intentionality have little to do with the punctum. As Barthes puts it, "to perceive the *punctum*, no analysis would be of any use to me."[47] Rather than something that can be planned by the photographer, the punctum accidently enters the scene and yet is somehow *not* present there, for what punctures one viewer about a photograph may not resonate with another viewer. Perhaps in the dark shadows beneath a subject's eyes, how their hands are positioned, or the red color of a bloodstain, the punctum lingers, even though we cannot articulate why it does so. Folding in Langer, we might say that the punctum expands in meaning as it persists through feeling—becoming fully articulate in its affective ineffability. This inarticulate semblance of the punctum is what takes hold of viewers, that continues to resonate long after their eyes have left the photograph, that promotes the conviction that they alone must grapple with this meaning. That is, the moment the punctum registers with viewers is not the end of the experience—the affect that pierces through in that moment does not dissipate but lingers. Because the punctum is known through feeling, rather than rational or discursive knowing, Barthes's semiotics is one that calls attention to the body, how bodies come to feeling and are moved by the more than discursive.[48]

Barthes's semiotics, then, confront us with how our bodies encounter, process, and store feeling and meaning that is not reducible to discursive, symbolic logic. In this way, the punctum "announces comportment," or confronts viewers with their corporeality and thus brings them to a "conscious experience of semiosis."[49] By bringing the affective corporeality of semiosis nearer, the punctum distances individuals from the impulse to interpret an experience quickly, easily, and discursively, if not challenging the idea that such discursive interpretation is even possible.[50] The punctum thus extends our understanding of Langer's felt life—felt life registers punctively, affectively, aesthetically. In turn, Langer's affective symbolism allows us to challenge whether Barthes's punctum need stay tethered to photography or artwork more generally. Remember that in Langer's post-Kantian aesthetics, art is not separate from but enmeshed in nature. Art's punctum is nature's punctum—both burst through an environment that is coded through discursive and more than discursive meaning, both connect the immaterial to the material, both, somehow, invite us to experience something that is objectively there (and not there).[51] Both call our attention to how phenomena move us—what elements register with our experiences, how meaning seems to emerge from a place we cannot articulate, how what compels our attention is broader than that of which we can make sense. Both emerge from evolutionary processes.

Unlike affect alone, nature's punctum helps to explain how something imposes on us, brings us to face the limits of discursive logic and know, instead, through feeling. Yet this punctive impulse to feeling is not a passive process in which the viewer does not participate. Because feeling, Langer explains, is not only something that happens *to* individuals; feeling is also an action, less a noun than a verb. As Langer argues most succinctly: "To feel is to do something, not to have something."[52] To make matters more complex, however, feeling as action does not suggest a conscious process. Feelings issue forth (read: individuals issue forth the act of feeling) subconsciously, in a process that occurs prior to and recursive with rational thought about those feelings. What lends feeling this ability to break through punctively is its immediacy, or as Langer resolves:

> It may seem strange that the most immediate experiences in our lives should be the least recognized, but there is a reason for this apparent paradox, and the reason is precisely their immediacy. They pass unrecorded because they are known without any symbolic mediation, and therefore without conceptual form.[53]

Feeling, then, moves into attention through affective meaning, that is, "without conceptual form," all the while staying attached to those material forms we encounter—say, a photograph or a landscape—that bring us to feeling. Such is how material forms come to carry and convey "felt life," a form of knowledge received intuitively, rather than discursively.

The Intuition of Beauty

Because Langer spent most of her career hoping to bridge the divides between sciences and the arts, she kept returning to the idea that humans encounter felt life in those moments of prescientific invention when discursive interpretation is stilled so that we can linger in feeling. Yet Langer would argue that the aesthetic resources of felt life are more than an epistemological tool for creativity; leading with prescientific invention is an evolutionary need. That is, intuition is a way of knowing that precedes and is recursive with discursive logic. To understand how intuition is a formative mode of knowledge creation, Langer turns to beauty.

Or, rather, Langer studies our reactions and engagements with artwork to understand the evolution of the human mind. "Art makes a prescientific but exact and detailed image of the mind," she would insist, because it is "ages of artistic practice [that] have developed what we may term man's prescientific knowledge of himself."[54] "Prescientific" here means knowledge that comes before biological or psychological concepts that explain how the body and brain work. With prescientific invention, Langer is interested in what can be known about feeling, symbols, and our evolutionary minds *through* feeling, through non-disciplinary ways of understanding. Langer's contemporary Thomas Kuhn would define this form of knowing precisely as "intuition," which he defines as a "typically subconscious mode of hypothesis formation."[55] Basically, intuition is how we know something before we know it. Similarly, Langer defines intuition as "the immediate apprehension of an object by the mind without the intervention of any reasoning process."[56] Important for Langer, however, is how intuition must necessarily underlie other forms of knowing as a faculty of mind that brings attention from one feeling, place, or concept to another:

> Intuition is, I think, the fundamental intellectual activity, which produces logical or semantical understanding. It comprises all acts of insight or recognition of formal properties, or relations, of significance, and of abstraction and exemplification. It is more primitive than *belief*, which is true or false. Intuition is not true or false, but simply present.[57]

Returning to the idea of felt life, we can say that intuition is the faculty that carries the mind from form to feeling, from symbol to song, instantly, punctively.[58] And the vivid values that so draw our attention—in their ability to issue feeling forth and puncture the faculty of intuition—take on an agential sense. No longer is a rational human agent the only interpreter of a static object; instead, the "object" issues forth its own vital aesthetic, which may be registered by human intuition in a way that is more than discursive.

This vital aesthetic, this more than discursive felt life, is the stuff of beauty for Langer. Social conceptions of beauty or highbrow notions of beauty at the hands of art critics are matters of taste, not beauty. When we are in the realm of the more than discursive, in the depths of feeling, when we are knowing intuitively rather than logically, then and only then, are we in the realm of beauty. In the realm of beauty, too, established concepts like subjectivity and objectivity must be reconsidered. For Langer, both subjectivity and objectivity help to explain the *feeling* of vital processes, rather than any sort of conceptual, cognitive intelligibility. Take how art can seem simultaneously public and private, paradoxically objective and subjective, symbolic of both cultural and personal feeling, Langer ruminates; such paradoxes need not be problematic if we "accept the tenet that art is an objective presentation, in symbolic projection, of the forms of subjective events; that its import is the nature of subjectivity, its production an act of objectification."[59] Rather than two different modes of thought, subjectivity and objectivity entail different registers through which individuals understand their world.

In this theory of mind, "'objective' and 'subjective' are definable before 'subject' and 'object,' namely as *ways of feeling vital processes*. 'Subjective' is what is felt as action; 'objective' is what is felt as impact."[60] The "subject" who does the feeling of the "object" occurs as a distinction after the fact.[61] As applied to the scientists to come, we can say that subjectivity plays an equally important role in scientific study as objectivity. Or, to put it another way, objectivity may be, at least in part, "prescientific," or caught up with and inextricable from that punctive feeling that reaches out to grab scientists' attention before they become "the scientist" studying a scientific "object." Of course, another objectivity emerges later in the process—in how scientists present their work—but such presentation first stems from this earlier affective encounter. This capacity to be moved punctively—perhaps by beauty as the biologists in this book will show—stems from intuition being our earliest and deepest teacher. Beauty teaches not what can be experienced through words but through feeling.

Toward a Method for Punctive Listening

Langer set out to change the scientific practices of her time by foregrounding ideas of affect and feeling as well as by focusing on the prescientific invention that precedes any intellectual practice. She would encourage scientists to return to that prescientific stage of discovery—therein to query their punctive, "felt life" encounters with nonhuman kin and their natural environments. As a by-product, scientists may come to a new understanding of objective observation as a relational process between scientist and environment, a process in which natural phenomena reach forth from the world punctively.

Langer's theories hold promise for present-day philosophies and rhetorics of science as they show that what is rhetorical about science exceeds, and possibly precedes, the bounds of text. What is rhetorical about science begins punctively, involves the felt life of the surrounding, environing world, and beckons scientists to bear witness. In other words, Langer asks us to consider feeling as an important part of scientific meaning making. She asks us to value what is not said but what is felt, to value the beauty that compels the scientist to pay attention in the first place. For rhetorics of science, Langer's work asks us to center aesthetics as primary to even objective methods of inquiry. For philosopher of science Adam Nocek, this shaping of science through aesthetics is precisely how Langer is most applicable to contemporary studies: her work allows for the development of research methods most "appropriate to the phenomena" under study, methods specifically tailored to and emergent from our subjects of investigation.[62]

To be clear, Langer's method of inquiry does not ask scientists (or artists or rhetoricians, for that matter) to "get beyond logic"; instead, she asks us to expand what we consider as logical thought. She asks us to lean into the intuition that is prior to and more expansive than our discursive ways of knowing. It is also important to know that *where* this method of inquiry happens was always important to Langer. Diffractively studying the arts and sciences brought her to study inquiry *in situ*: in the studio and in the laboratory. Theory, Langer believed, emerged in the middle of the making, not through the stop-and-stare habits at galleries and museums, or in published articles.[63] The prescientific work of invention is there, in the studio, in the laboratory, or even prior to the laboratory in the field. It is there—where finding beauty in what is being studied is enough—that punctive, felt life can break through our thought processes, where affective ruptures can guide us to think, know, and feel more than discursively.

In what follows, interviews with evolutionary biologists will challenge and extend the idea of "nature's punctum," as each thinker offers different attributes of punctive listening in their natural environments. Biologists, Langer would say, are the perfect case studies for this work because "'life' is obviously not easy to define." She insists, "No one [knows] as surely as the biologist" the difficulty of drawing a line between animate and inanimate life.[64] Such is why this book sets out to observe how these evolutionary biologists come to study some of the most beautiful forms of life in nature, the unexpected thread of which is that each biologist in the pages to come has some story about nontraditional or non-disciplinary ways of knowing paving the way for their work. Their experiences will speak to how biologists use sensory, bodily, rhetorical practices to quiet their interpretative instincts and traffic in the more than discursive symbolizing of natural environments. If these experiences are facilitated by listening to nature's punctum, then we will come to know the punctum as an affective jolt that pierces the gut, the skin, the ear as readily as the eye.

2

Punctive Listening

It is this way with wonder: it takes a bit of patience, and *it takes years putting yourself in the right place at the right time*. It requires that we be curious enough to forgo our small distractions in order to find the world.

AIMEE NEZHUKUMATATHIL, *World of Wonders*[1]

I am, by a flood, borne back to that wondrous period, ere time itself can be said to have begun; for time began with man. Here Saturn's grey chaos rolls over me . . . and in all the 25,000 miles of this world's circumference, not an inhabitable hand's breadth of land is visible. Then the whole world was the whale's. . . . Who can show a pedigree like Leviathan? . . . I am horror-struck at this antemosaic, unsourced existence of the unspeakable terrors of the whale, which, having been before all time, must need exist after all humane ages are over.

HERMAN MELVILLE, *Moby-Dick*[2]

Evolutionarily, we share a common ancestry with whales, our relatives who filled the ocean with song tens of millions of years before human animals came onto the stage.[3] Of course, for Melville, Leviathan was anything but a musician. Leviathan was ancient, antemosaic, and—Melville was sure—certain to live after us, considering that whales seemed a death sentence to anyone who gave chase. Indeed, one thing Melville never attributed to Leviathan was song. Nevertheless, this particular passage of *Moby-Dick* may set the stage for human listeners to open their ears to the music of nature. Here, "Saturn's grey chaos" roared across the cosmic chronos. Now, a quick trip to YouTube allows us to hear the rings of the planet Saturn as they hum and creak in their cyclical revolution around the cold orb, like a subway train approaching its station— the sound of ice crystals and sand intermixing.[4] In the passage, too, whales gently rolled across the abyss, filling the circumference of the world over with their spouts of breath and breaching crashes. Now hydrophone recordings enable us to hear their deep moans, guttural clicks, and screeches reminiscent of laughter as they reverberate hundreds of miles across the deep.

The whir of Saturn's rings and the echoes of the whale's cry fill the world with sound from the unlikeliest of places, each "song" covering a different end of the universe's dark depths.[5] And at least the songs of the humpback whale follow the rhyme, rhythm, and phrasing patterns similar to human musical compositions, a resonance that brought biologist Roger Payne to wonder whether these sounds derive from a shared ancestor "tens if not hundreds of millions of years old," an ancestry that would make the melodies that sometimes spill from our lungs "thousands of times older than we are."[6] And if that's the case, Roger Payne asks, "Is it possible that the universe sings?"

Traditional scientific methods offer biologists no way to answer such a question—whales and song do not quite fit the bill for subject under study in a laboratory. Even the first report published in *Science* in 1971 about the songs of the humpback whale introduced the phenomenon as "a series of *surprisingly beautiful* sounds."[7] That scientific attention was here directed by beauty is an occurrence that Langer may have considered an instance of "prescientific invention," a moment in which attention is directed by feeling, here by the beauty of the songs of the humpback whale. What this case study offers—a case in which scientific study cannot be so neatly contained in the lab—is how biologists might attune themselves to nonhuman ways of knowing in natural environments and how engaging these "surprisingly" aesthetic moments may move biologists beyond discursive semiotic modes and into feeling nature's punctum.

Since the 1970s, Roger Payne had written extensively about whale songs, and in his recollections, stories of Katy Payne, his spouse at the time, play a key yet subtle role.[8] While Roger Payne spent his days observing whales, Katy Payne recalls spending her days "chasing children."[9] What Roger Payne did not narrate and what Katy Payne recalled in our interview is that as she "chased children," she pored over spectrograms in her spare moments to view the structure of whale songs and played the recordings of humpback whales on repeat throughout the house.[10] Through the sustained practice of intensive, ambient listening over many years, Payne began to realize that these sounds were repeating, rhyming, and changing in relation to one another.[11] In short, it was Payne who performed deep analysis of the recordings, came to consider the whales as composers, and determined that whales were not just sounding, but singing in response to one another.

Payne's contributions to the scientific annals, to be detailed throughout this chapter, would come to be defined as the "sensitive" work needed to decipher the songs of the humpback whale.[12] Not often theorized beyond a gendered way of knowing, Payne's "sensitive" approach is a fully embodied mode of listening to the affective force of impositions from natural environments, defined in the previous chapter as "nature's punctum." Langer theorized that something like "felt life" compels and directs our ways of knowing prior to (and/or separate from) disciplinary reason, but she was unable to direct scientists in *how* they may allow feeling to lead the way in inquiry. Payne, then, practices what Langer may have never imagined: how to observe felt life in natural environments, how to listen to nature's punctum. As will become clear, Payne embodies and demonstrates a method of "punctive listening" exemplified by the other biologists in this book.

Punctive listening comes before, or apart from, disciplinary epistemologies, it would seem and, Payne would say, is less a method of inquiry than a way of being in the world as an observer. Of course, what it means to be a scientific "observer" is also up for redefinition here. For now, it is enough to say that Payne offers a method of "punctive listening" for biologists, a method that prioritizes feeling as reason and is manifested through two distinct practices: (1) what she calls a practice of "simple observation," which has less to do with visual ways of seeing and more to do with an adherence to the semiotics of the nonhuman world that necessitates a withholding of human symbolic interpretation, and (2) a certain bodily comportment in the field in which the observer is humbled by the beings before them. Payne's method of punctive listening offers a form of empiricism in which rhetorical resonance with subjects is prioritized over discursive, rational knowledge about those subjects, an empiricism in which scientists become more like witnesses than observers.

In the context of this book, Payne offers contemplative practices that promote deep listening to her surroundings and the beings therein over hypothesis-driven research. I should also note that Payne was the first person I interviewed for this project, and unbeknownst to me at the onset, every other researcher I would go on to interview had some story about the influence of Payne on their work, stories I promise to relay in due time. Suffice it to say that if we are to challenge scientific ways of knowing, I am now convinced, we must first sit and listen with Katy Payne.

Hearing Leviathan Sing

When Katharine "Katy" Boynton Payne invited me into her living room, instructed me to sit, and began to play the songs of the humpback whale over surround-sound speakers, I was confronted with sounds that have been known to move people to tears. Payne, along with former research biologists Roger Payne and Scott McVay, would bring these songs to public attention in 1970 with the release of the vinyl album *Songs of the Humpback Whale*. The album, released prior to scientific research on whale songs, would sell over thirty million copies, becoming the best-selling environmental album in history and attracting the attention of musicians like Pete Seeger.[13] Further, the album's impact would spark international whale conservation efforts as well as contribute to the organization of Greenpeace. Musician and philosopher of music David Rothenberg marks the album as one that made human animals stand at attention: "When [we] first heard these sounds, our sense of whales suddenly changed. It is the song of the humpback whale that made us take notice and care about these animals—the largest that have ever lived on Earth. We would never have been inspired to try and save the whale without being touched by its song."[14] No exaggeration, then, to say that whale songs changed how humans understood the chorus of the sea.

Roger Payne, however, did not always study whales. Instead, he studied echolocation in bats as well as owls' ability to localize sound due to the asymmetry of their skulls. His focus would shift in the late 1960s, the year he read *Moby-Dick*. The literary origins of his scientific work are not ones that he remembered. Instead, it is Katy Payne who remembers how that book shifted their life's work. "He wouldn't say it started with [*Moby-Dick*], but I remember that it started when he was reading that," Payne recalls. "And he just . . . swallowed [the book]."[15] One particular passage on conservation, Payne reports, was especially memorable. Near the end of the novel, Melville begins to doubt whether the seemingly eternal Leviathan can withstand the onset of the whaling industry and humanity's brute force. As Melville writes:

> The moot point is, whether Leviathan can long endure so wide a chase, and so remorseless, a havoc; whether he must not at last be exterminated from the waters, and the last whale, like the last man, smoke his last pipe, and then himself evaporate in the final puff.[16]

Roger Payne ruminated on the passage for quite some time and finally exclaimed, "Well, nobody knows anything about whales. Let's go find out."[17]

In 1967, the two would venture into the Atlantic to listen to the recordings of US Navy engineer Frank Watlington. In 1971, Roger Payne would publish "Songs of Humpback Whales" with Scott McVay, a researcher who had studied the cognitive minds of dolphins under the tutelage of John Lilly.[18] Before their article, which graced the cover of *Science*, scientists were aware that whales made "short broad-band clicks, longer narrow-band squeals, and complex sounds."[19] Roger Payne and McVay were the first to show that these sounds were actually composed as songs, presenting in *Science* the entirety of a thirty-minute whale song in which they identify fixed patterns and repetitive phrases therein.[20]

The Paynes and McVay met unique challenges in studying the songs of the humpback whale because these songs bring biologists to the limits of scientific experimental methods. Without the ability to study their subjects in a laboratory setting, their only option was to pursue and observe the whales by boat, record their songs with a hydrophone, and create visual representations of the songs with spectrograms. Spectrograms, pictured on the *Science* cover,

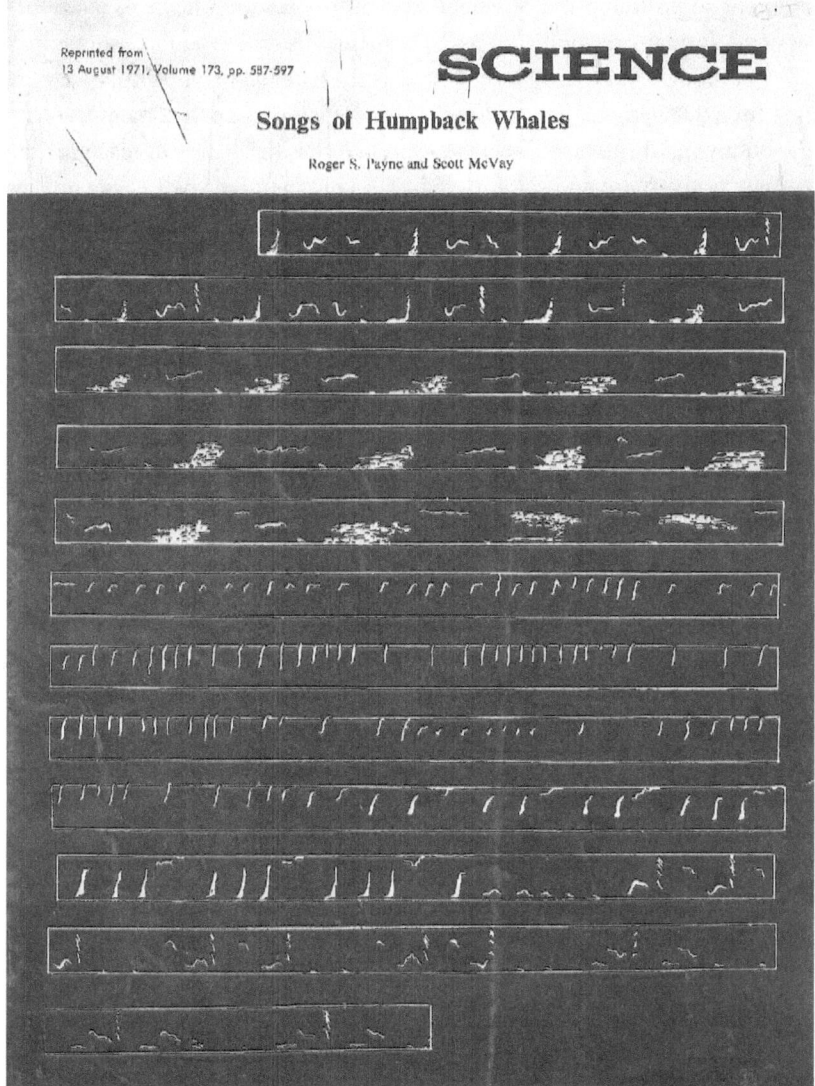

FIGURE 2. Spectrograms of whale songs on the cover of *Science*. Roger S. Payne and Scott McVay, "Songs of Humpback Whales: Humpbacks Emit Sounds in Long, Predictable Patterns Ranging Over Frequencies Audible to Humans," *Science* 173, no. 3997 (August 13, 1971): 587–97. Reprinted with permission from AAAS.

are the visual representation of a signal's frequencies over time—in this case, the pattern of a whale song. As Roger Payne recalled, "[Studying whales] is a lot like astronomy... you can never perform an experiment but must wait for nature to present you with something interesting to observe."[21] The moment the Paynes first listened to the songs of the humpback whale would mark a moment so profound that it would spur on the research fields of bioacoustics and animal communication. Yet somehow Katy Payne has no memory of what thoughts or questions emerged as she listened to whale songs for the first time. Or perhaps I should say that there is no discursive memory, only the bodily mark she performs when she stretches her hands to reach for the corners of the room and sighs, "Ahhhh." Awe. When I pushed Payne further to articulate her method of inquiry, to name what it feels like to hear whales sing, she said, "I can. But I'm going to make you do it."[22]

Listening with Katy Payne

Prior to this moment, we had been seated at the dining room table across the house, talking about Payne's research. Payne was candid about most things, such as her time as an undergraduate at Cornell and the chaotic nature of raising four children in Patagonia. But she became withdrawn when I asked about her first memories hearing the songs of the humpback whale. Her method seemed so very different from more disciplinary methods of study or my conception of scientific method, and I wondered what sensations accompanied that original feeling of awe. I wondered, that is, how Payne had come to make sense of her experience. But she hesitated and veered away from the topic. I accepted the digression, listening as she discussed being a music major at Cornell, meeting Roger Payne through chorus, and studying the social behavior of ants with E. O. Wilson. Then I asked her to return to that moment when her life's work shifted, when she went from being a student of biology to becoming an expert bioacoustic researcher.

It was fifty-two years ago in the waters surrounding Bermuda; she would have been thirty years old, I remarked, in the hopes that locating a certain age and time might spark memories Payne had not previously been able to recall. Perhaps a question or two emerged when she first heard these strange sounds, which she and Roger Payne did not then know as songs. "No," Payne would insist, "I don't come at life with preformed scientific questions."[23] We let quietness linger for a few seconds, and then she sighed; I was missing her entire point. That is when she led me into the next room and asked me to listen. Slowly, she turned the volume knob so that the echoes of the humpback whale reverberated across the room until the pens in a glass jar on her desk shook a little.

The reason that people are moved to tears when they first hear the songs of the humpback whale, Roger Payne believed, stems from how these songs allow individuals to identify with their evolutionary ancestors, with "something unaccountably ancient overmastering [us]."[24] Rothenberg studied emotional reactions to whale songs, with one of his research participants marveling at how these sounds were not like ones usually associated with large mammals: "It wasn't primal; it wasn't animal-like." Another remarked: "It was an Ur-music; almost a religious experience."[25] I did not cry or feel deeply connected to Gaia in that moment. I simply felt strange. These sounds were profound, and—Roger Payne was right—I did feel an identification with them, just not one that I understood, exactly, or could articulate. Perhaps that feeling was one of being connected to ancient, evolutionary origins, as is hypothesized. I certainly feel that now, after making a playlist of whale songs and listening to them for hours on end (for as many hours as it took me to read *Moby-Dick*, pore over bioacoustics research, and write and revise this chapter). But recalling that moment, I only remember the strangeness.[26] After that, discursive symbolizing took over: here was a world-renowned researcher playing the recordings on which she based her life's work. I had better have something to say when she pressed "stop."

Replaying the recording of our interview, I am reminded that Payne and I talked off and on during the procession of the whales' songs. I would ask clarifying questions: "Is this one whale or two?" Payne would answer: "One. One male whale." Then she would direct my attention to some aspect of the sound: "Do you hear how it echoes? . . . The echoes are from the bottoms of the waves and the bottom of the sea." I listened for a minute more, then asked how far these echoes reached. "Hundreds of miles, especially the low sounds," she answered.[27] Something about this magnitude—a song reaching for hundreds of miles—changed how I heard the short, high-pitched clicks. "There seems to be a playfulness to it, doesn't there?" I prompted. I saw her give a moment of pause, but then return a smile. "Doesn't it feel that way?" she encouraged. We listened in silence for several more minutes. Once again, I broke through the echoes: "I don't know if 'longing' is the right word, but there's a sense of—I don't know—almost like a reaching. It just seems like . . . they're reaching for something or someone else." Her smile brightened because I was connecting to their music. I was, however, also on the precipice of making their music my own, prompting Payne to hypothesize: "I think you may be anthropomorphizing."[28] The elongated notes have to do with the medium of water and how much longer it takes sounds to travel through water. My ascribing "longing" to that occurrence was letting my emotion get in the way—to assign human emotions to these large mammals risked appropriating their song, Payne implied. Jumping to conclusions about how the whales use their songs meant that I was not really listening to the song itself.

Although the original move to name these sounds "songs" may itself seem anthropomorphic, Payne is careful not to suggest that biologists know what these songs mean, at least according to rational interpretation. Certainly, these sounds do mean—as Payne would come to describe whales as "composers" who craft songs full of meaning, full of intention, but meaning fully articulate in its sound. That is, these songs mean affectively, not discursively. To use Langer's terms, these songs symbolize like artwork, through feeling. Listening to them—*really* listening, not just hearing—as Payne will show, requires what poet Aimee Nezhukumatathil describes in the opening epigraph: patience, curiosity, and years of "putting yourself in the right place at the right time."[29]

Part of being in the right place to hear whale songs for Payne was finding them "out on a little boat . . . clinging to a cork" in the ocean.[30] Even with hydrophones submerged and headphones adorned, there is still no certainty of scientific experiment, no certainty of whales' appearance, much less their song. And because the ocean is such a noisy—or as Payne says, "rowdy"—environment, she easily could have considered these sounds part of the cacophony of the sea with whales singing alongside snapping shrimp and boat motors. Only after listening to recordings of the humpback whale thousands of times over did she, Roger Payne, and McVay realize that, following the characteristics of birdsong, humpback whales compose "fixed patterns of sounds that are repeated" and thus indicative of song.[31] These songs last anywhere from seven to thirty minutes, and their sounds are low and long, far-reaching notes mixed with high-pitched clicks and squeals. Payne cannot say whether the whales feel the longing of distance and separation, as the echoes sometimes make it seem, or whether they are pleased with their compositions. She wonders about such matters but is sure of only one thing: individual whales change their songs in response to others around them.

As the singing season, which overlaps with the feeding/mating season, progresses, the songs of individual whales shift slightly, with certain notes flattening and then eventually dropping out. Other whales will detect and often play with that composition change, imitating it and sometimes altering it with their own inflection.[32] The changes to these songs slowly travel across the ocean so that "all whales in the same ocean sing basically the same song even though as changes get made on one side of an ocean they may take more or less time to transfer to the singers on the other side."[33] And the songs undergo such rapid change throughout the season that by the start of the next year, the song is almost entirely different. "Like improvisation in human music," Payne writes, "changes seem to be generated by an internal process, and as in music, the imitation that then occurs reveals listening and learning."[34]

Considering whales as intentional composers, rather than mere noisemakers, brings contemporary marine biologists like Hal Whitehead and Luke

Rendell to adopt the hypothesis that cetacean behavior is culturally determined. Whitehead and Rendell argue that if humpback whales sang one or two stereotypical songs, then perhaps their tune could be considered genetic.[35] Instead, these songs change rapidly across oceans, a "cultural attribute" that has "no known parallels outside humans."[36] Extending Whitehead's cetacean culture theory, biologist Ellen Garland is working to cull data from researchers across the globe to construct a map of how humpback whale songs travel horizontally across oceans.[37] The stakes of such a map, her team believes, will confirm that humpback whale song is a product of cultural evolution.

Largely, however, the songs of the humpback whale remain as much a mystery to biologists now as they did fifty years ago when Katy Payne, Roger Payne, and Scott McVay first reported on them. In the years since the Paynes released *Songs of the Humpback Whale*, researchers have determined that while both sexes issue calls, only males sing.[38] The leading idea, then, is that the songs have something to do with sexual selection, and because male humpback whales improvise and tweak each season's particular song, Payne contends that there may be some evolutionary advantage to improvisation to which females are attracted.[39] In an overview of research on humpback whale songs, marine biologist Louis M. Herman presents three competing propositions about the function of whale song: "to attract females to individual singers, to mediate interactions between males, and to attract females to a lekking [courtship] area."[40] Undoubtedly, Herman argues, these songs serve multiple functions. One particular whale call, known as the "whup," may serve as a greeting or "hello," suggests marine ecologist Michelle Fournet, whose work is featured in chapter 4 in more detail. Through an ongoing study, Fournet hopes to prove that the "whup" may be personalized to announce each individual's presence: "Hello, I am [so-and-so]."[41] Yet Whitehead and Rendell clarify that little about studying humpback whale culture is as straightforward as is hoped.[42]

The mystery of whale song and calls has something to do with the whales' size but more to do with the vastness of the ocean. "Half of the experience" studying whale songs, Payne reports, "is wallowing around in a boat." Again she insists: "You've got to go hear them for yourself, and you've got to sense the ocean yourself and know how huge it is and have a feeling for what's going on."[43] When faced with the vastness of the sea, it seems, we are less inclined to anthropomorphize, to make those songs or the depths of the sea mean so discursively. Payne explains the study of whale songs as one that requires sustained attention: if you are met with a song, you must wait to confirm just who is singing. Then you wait for a sign of breath to shoot upward into the sky. "You're using your eyes, even though your eyes are the worst guide,"

FIGURE 3. Whale watching (and listening) featuring Katy Payne and Annie Lewandowski. Photo courtesy of C. Gabriele, Hawaii Marine Mammal Consortium.

Payne remarks. "If you're listening to a song—and it suddenly becomes rather soft . . . all eyes. All eyes to all angles of the horizon because it means this whale is coming up for a blow. Then you have to wait another twenty more minutes [until they come up again]. And you get a little bit closer, a little bit closer."[44] Such listening cannot be rushed, and there is no mistaking the power dynamic: the human symbolizing animal is quite insignificant, at the whim of whale and sea. Studying the songs of the humpback whale, then, requires not only a certain mode of inquiry, but also a way of being with them, a way of allowing whales to speak—well, sing—for themselves.

Simple Observation

Payne is not big on definitions, but there is one definition about which she is certain: in her estimation, she is not a scientist. At least she kept clarifying as much during our hours-long interview. Each time I referenced her as a "scientist," she explained that she had no such credentials or formal training. Instead, Payne insisted that she "simply was an observer," and observing was the greatest privilege that she knew.[45] Payne may have begun her studies in biology as an undergraduate at Cornell, but "scientists," she learned, had markedly different ways of approaching natural phenomena. Having spent her childhood climbing gorges in the rural countryside of upstate New York, Payne felt ill at ease studying "life" from "the molecular orientation of biology"—that is, on the scale of chemicals, genes, and other elements invisible to the eye.[46] By contrast, Payne's way of observing her natural environment was "noticing things and enjoying them . . . noticing whole organisms and societies."[47] Payne would thus turn away from biology and value instead the knowledge accrued from her lived experiences. Subsequently, she opted to major in what she considered a more creative outlet—music—which would eventually inform how she listened to the songs of humpback whale.

Payne's training in music and her way of being in the world more generally, of withholding disciplinary and discursive tendencies, have altered the field of bioacoustics. First, of course, Payne explained that whale songs are evolving, cyclical, and inundated with meaning. Decades later, she revealed that elephants communicated infrasonically—the focus of the following chapter—which she discovered after sitting and meditating outside the elephant exhibit at the Washington Park Zoo in Portland. Payne detected slight vibrations in her chest when she approached the elephants, feeling as though the lowest key on the organ had just been struck. Payne's process of interpretation occurs alongside other beings in the world, with her training in music teaching her how to understand that sounds mean affectively more than discursively and thus require attention from the whole body.

Such practices stand to challenge notions of the degree to which scientific practice must be "objective" to hold epistemic authority. It is not as though Payne depended upon these subjective practices because she had no disciplinary training. As will soon become clear, Payne went to great lengths to maintain this certain way of relating to the world, by leading with punctive listening before applying traditional disciplinary methods. Yet contrary to my hypothesis prior to our first meeting, Payne does not completely dismiss notions of objectivity either. Punctive listening, then, retains objectivity insofar as that objectivity is diffracted through aesthetics and more than discursive ways of knowing—in this case, music. Punctive listening is a method of empirical inquiry in which rhetorical resonance with subjects is prioritized over discursive knowledge about those subjects, in which meaning derives quite literally in a different sense, a sense in which feeling is fully articulate.

Here, we begin to see Payne's resonances with Susanne Langer, who had a macroscopic definition of the symbol that brought her to theorize how more than discursive modes were already fully articulate in their affective meaning. Nature's punctum need not be translated into discursive language to convey knowledge or become intelligible, according to Langer—beauty conveys meaning on its own accord. Langer, we know, was much like Payne, in that she was also a musician, writing as much about music in her work on feeling and form as about visual art, resisting disciplinary bounds at every turn.[48] Other musicians like Rothenberg and Seeger have composed music alongside whales, understanding their songs through human art forms. Many contemporary scientists, on the other hand, produce research on whales only by studying spectrograms in the lab. Payne finds either tactic too reductive. As she puts it: "The whales singing are to me neither science nor art. . . . I can apply scientific rules to what I'm hearing, but then I'm leaving out half of what I'm hearing."[49] It may be most fitting to say that Payne's background in music brought her to expansive definitions of rhetoric in which meaning is always diffracted, in which researchers may use disciplinary tools to hear, but in which a full-bodied, simple observation of beauty is needed to really listen.

Listening may, indeed, be the most direct way to access affect, emotion, and nature's punctum, as sound communicates directly to the emotional centers of the brain. According to audiologist and sound engineer for the Cornell Lab of Ornithology, Bill McQuay, it takes a while in visual processing for information to reach the visual cortex in the brain; sound is more immediate.[50] McQuay gleaned this information from neuroscientist Seth Horowitz, who joined McQuay's NPR podcast to report how sound shaped the human brain's evolution: "You hear anywhere from 20 to 100 times faster than you see, so that everything you perceive with your ear is coloring every other perception you have, and every conscious thought you have. [Sound] gets in so fast that it modifies all other input and sets the stage for it."[51] Hearing is also more panoramic than vision; we hear in 360 degrees. This is how sound "ties" us to our environments, McQuay posits—evolutionarily, our ears had to be sensitive enough to wake us from sleep should something stir in the trees as we slumbered. This, too, is why sound is "felt"—sound delivers an immediate bodily response so that we can quickly avoid danger. Sound routes to "pre-cortical" areas of the brain, or those areas of the brain that deal with basic functions rather than conscious thought, and according to neuroscientists like Horowitz, emotion is one such basic function triggered by sound. "Emotions are rapid delivery systems in the brain," Horowitz reports, "and sound drives emotion."[52] Sound's ability to reach our depths is also why when Langer talks about "prescientific invention," she talks about feeling: if we want scientists to know beyond disciplinary reason, to know through feeling, then we must teach them to listen. Just as listening—the act of hearing sound and resonating with it, feeling it, knowing it—is more than hearing those notes we consider as "sound," simple observation, too, is not directed narrowly to a visual process—it is an immersive, 360-degree experience.

Other scientists rarely take an approach like Payne's because—quite frankly—it is inefficient. Payne identifies haste as one of the main issues with the American education system: "This is the problem with higher education. There are reasons why those involved in higher education try to guide their students and their listeners to certain interpretations: because they're saving a lot of time. All of this has been done and works, and all that's been done and doesn't work. And . . . so I just say: be wary."[53] Rather than working with established interpretations and testing hypotheses, Payne is more interested in this "simple" observation. "Simple" is a word she often uses in speaking of her way of being in relation to natural phenomena: "I *simply* was an observer. Both as a scientist and as an artist."[54] This claim to simplicity, like her resistance to expertise, is itself a rhetorical move, one that ensures humility and the adherence to beauty as a method of praxis. If she went in as "the expert," Payne may not take the time to punctively listen. In short, her "simple observation" ensures that Payne never looks for anything; she never approaches her natural environments with specific questions and the need to find certain answers (or certain kinds of answers). Instead, she waits for what resonates, bodily, affectively, sensorily; she dwells with her subjects of study; and she relays only that experience to others.

Payne's way of listening, then, has less to do with hearing what can be conveyed discursively later and more to do with what Langer called "felt life." That is, Payne's simple observation allows her to feel the evolutionary connections between her own being and the vitality flowing through the other beings in her environment. Life does not subsist, unnoticed, by Payne; instead, she observes her surroundings carefully and listens deeply to what she does not know. Eventually, and understandably, one central question did emerge for Payne and even begins to animate her observations: "What does it feel like to be that animal?" Yet even stating that question during the interview felt too specific for Payne, who quickly rephrased the question more broadly: "What is it?" Even more quickly than her rephrasing was the qualification that came almost in the same breath: "[But] I wasn't really a *scientific* student. I was just interested in whatever we could learn."[55] In this way, Payne's way of listening is more inductive than deductive: with no notions of a symbolic hypothesis to prove or disprove and resisting the idea that the human can "discover" anything, Payne sees it as her only task to notice, to attend to the songs of the humpback whale, and the other punctive impositions that spring forth from the natural world, to traffic in what anthropologist and new materialist theorist Eduardo Kohn has called an "open" semiosis.

Listening to an Open Semiosis

If humans are part of that nature they seek to understand and thus "mind" emerged from nature, as Langer argued, then there must be a shared semiosis among human and nonhuman kin, a semiosis that is "open" to the broader, natural environment, to other ways of knowing, to meaning beyond the discursive symbol. Langer had theorized *that* semiosis extended beyond the human, and Whitehead before her placed semiosis deep in sense perception, but neither specified how human communication was conversant with and reliant on nonhuman animal communication. Such posthuman concerns would be addressed much later by Kohn in his studies with the Runa of Ecuador's Upper Amazon. Kohn's posthuman semiotics provide an explanation of how humans wade in semiosis beyond the symbol, because symbols are reliant upon, emergent from, and altered by nonhuman ways of representation.

Deriving in large part from Charles Sanders Peirce's distinction between icon, index, and symbol, Kohn posits that symbolic thought is not the only way of communicating and making meaning for humans. Indeed, symbols—which are characters or words that represent a concept, process, or object indirectly—are but one kind of sign. Icons are signs that share a likeness with what they represent, and thus are recognized across cultures, such as someone in the forest saying "tap tap" to convey the image of chopping down a tree.[56] Indexes are signs that point to what they represent but do not share a one-to-one likeness with that thing, such as the crash of a tree in the forest startling humans and nonhuman animals alike. The crash is not the fallen tree, but points to the idea of it (i.e., a tree has fallen).[57] Whereas symbols are reliant on abstract, arbitrary thought processes and thus relegated to humans for Kohn, icons and indexes are shared across species. The open semiosis, then, describes how humans are "open" to broader ways of signing and thus making meaning beyond symbols, to ways of thinking that stem from evolution and ecology itself.[58]

A communicative resource, discursive symbols nevertheless separate humans from broader ways of communicating in the open semiosis because they ask us to make sense of the world from a detached perspective.[59] As Kohn contends, symbolic thought has "the propensity . . . to jump out of the broader semiotic field from which it emerges, separating us, in the process, from the world around us."[60] By way of explication, Kohn details one instance in which he experienced an anxiety attack while on the road back to the region of Ávila. His thoughts were interior, rapid, and detached from the surrounding world. This detachment dissipated—apart from his directed intention—when a tanager appeared through his binoculars. Then, Kohn "experienced a sudden shift" in which his "sense of separation simply dissolved. And, like the tanager coming into focus, I snapped back into the world of life."[61] Kohn references this moment as one of "regrounding," a returning to the larger semiotic field always present and yet perhaps rarely noticed. Whereas rhetoric scholars such as Diane Davis may now acknowledge that "human beings are not the only symbol-using animals"—and the complex composition of whale songs seems to support that hypothesis—Davis herself asks, "How might [extending rhetoric beyond the human] transform our relation to and sense of responsibility for the rhetorical traditions we have inherited?"[62] As Kohn shows, the symbol, or the makings thereof, is not all that we have inherited. Because symbols are open to and constitutive with icons and indexes, the language and song relegated to the human come from a broader semiosis that our evolutionary ancestors have spent centuries perfecting. Similarly, Parrish reminds us that because human animals are not the first rhetors, neither may we be the best.[63]

Not only is semiosis a more than human process, Kohn contends, but it is how life itself unfurls. Life is—from the ground up—a product of sign processes.[64] "Semiosis," as Kohn defines it, "is the name for this living sign process through which one thought gives rise to another, which in turn gives rise to another, and so on, into the potential future."[65] In this way, signs are not things or ideas in the mind, but representational processes that attach to words, scenes, and sounds. Whereas many scholars consider signs a product of the mind, Kohn would reverse the direction. "Signs don't come from the mind," he clarifies; "rather . . . . what we call mind, or self, is a product of semiosis."[66] Scientific findings echo Kohn's premise when studies show that "animals think a lot like humans; primarily through intuition, instinct, and yes, emotion but also sometimes through logic."[67] In other words, human language processes are not all that different from the processes by which life—nonhuman animals, forests, even evolutionary ecology itself—communicates, affects, unfolds. Through this idea of the open semiosis, Kohn argues that icons and indexes are resources that may allow human animals to conceive of meaning beyond the symbol, and that, in this way, human animals may listen to other species.[68] As an example of how humans make meaning in the open semiosis, Kohn details the experience of two Runa women who hear their dogs switch from barking to yelping in the presence of a predator. Because both the women and the dogs are beings in the world who share understandings of icons, indexes, and images, the women understand that the dogs misinterpreted the look and size of a mountain lion for a deer. Because the Runa women share the open semiosis with the dogs, they are thus able to interpret "how [the dogs] came to understand the world around them."[69]

Langer, of course, would have challenged Kohn's definition of the symbol, in that in his semiotic work, icons and indexes do not seep into the bounds of discursive symbolizing. Langer's multifaceted symbology, by contrast, included discursive symbols as well as affective, or more than discursive, symbols. In this way, Kohn's and Langer's theories run parallel to one another with differences only in terminology. Langer and Kohn would have agreed that semiosis is open to the surrounding, natural world, but Langer's emphasis on "felt life," on that punctive pull to feeling, best articulates how humans traffic across this semiosis, or slip from one way of knowing to another, to a broader way of "simple" observation. What Kohn does not fully explicate, then, is how humans come to disrupt their primary, symbolic ways of making meaning in the first place. As Payne's listening to whale songs suggests, some rupture—some more than discursive appeal to the senses that cannot be interpreted easily—may be what punctures our rote ways of knowing. As defined in the previous chapter, nature's punctum is that asignifying rupture from the natural world that cannot be orchestrated or planned. Barthes theorized that the punctum could be seen; through Langer, we theorized that the punctum allows for a sense of "felt life"; and here—through Payne—we understand that the punctum can perhaps best be known and felt through listening.

Punctive listening calls attention to the affective, sensory, and bodily work of semiosis. If the open semiosis is known through feeling, rather than rational or discursive knowing, then nature's punctum confronts human animals with how their bodies encounter, process, and store rhetoric that is not reducible to or categorized exclusively by discourse. The example of sighting the tanager, however, suggests that, for Kohn, the open semiosis is conceived imagistically. *How Forests Think* even concludes with this succinct statement about what forests teach us: how to think with images. "Images amplify," Kohn argues, "and thus render apparent, something about the human via that which lies beyond the human."[70] But if forests think in images, oceans—those dark depths marked by the lack of visibility—must think in sound.

Indeed, Payne's method of listening, I argue, moves us closer to an explanation of how to listen to the open semiosis. As she remains always available to listen for moments of asignifying rupture, Payne suspends discursive interpretation to avoid and disrupt any dualistic divide between herself and her subject of study. In semiotic encounters, meaning is traditionally thought of as being produced or received, but in Payne's praxis there is no clear sense of where her meaning begins and the whales' meaning ends. And for those "grounded," like Payne, in the open semiosis, sound and image are not so separate—both are bodily experiences that allow individuals to receive more than what can be heard or seen. The idea of "listening," then, to the open semiosis articulates how Kohn's theory may be expanded beyond the visual to the affective and defines how researchers can use sensory, bodily, rhetorical practices to quiet their interpretative instincts and traffic in the more than discursive rhetoric of natural environments. If more than discursive ruptures occur when the brain cannot make sense of what is before it, such a stop to interpretation need not be tied so firmly to visual images.

Music, after all, is closely connected to our emotions and is unable to find an exact correlative in discourse. When we talk about music, we bring sound into the discursive symbolic realm, explains musicologist Nina Sun Eidsheim, but even through our discourse about music, "the presymbolic does not cease to exist."[71] When Eidsheim points to how music is able to influence listeners through a "presymbolic domain," she identifies what Langer would have called affective or "presentational" symbols. The songs of the humpback whale move us, influence us, and can be understood by us at the level of the presymbolic or more than discursive. Accessing meaning on that register requires understanding that listening happens primarily through the body than through discourse.[72] Eidsheim turns to research on prenatal hearing as evidence for this occurrence: scientific research confirms that reactive listening occurs as early as sixteen weeks into fetal development, or eight weeks before the ear is fully formed. Because listening occurs prior to the formation of the ear, Eidsheim concludes that "hearing begins in the skin and skeletal network ... begins with the rest of the body in a 'primal listening system.'"[73] In a word, just as image need not be tethered to the eye, sound need not be tethered to the ear—listening is a full-bodied experience.

Bodily Comportment in the Field

Decades after her departure from biology at Cornell, Payne would reevaluate her stance that the discipline suffered as a result of its "molecular orientation."

She would encounter research on the chemical defense of beetles by Thomas Eisner and on the social life of honeybees by Thomas Seeley and conclude that systematic training need not be mutually exclusive from her method of listening. "See this is ideal," Payne would remark, "know the canon. There's where I fault myself." Know it, she would insist, and then "expand it."[74] But at least from my vantage point, it seemed clear that Payne was expanding the canon precisely because she gave herself permission to work around traditional scientific methods.

For the same reason, Payne's scholarly mentor, Charles Walcott, then director of the Cornell Lab of Ornithology, dissuaded Payne from returning to graduate school when she expressed interest in pursuing a graduate degree. After the Paynes separated in 1985, it was Walcott who suggested that she return to the lab in Ithaca and continue her work. There, Payne would spend several additional years studying the songs of the humpback whale and publishing her findings on rhyme, repetition, and mnemonic devices.[75] But before she felt confident in this work, Payne compared her skills to those of the graduate students in Walcott's lab and became convinced that she was falling short. Finally, she told Walcott that his students "know all kinds of things I need to know," and asked whether Walcott would sponsor her through Cornell's graduate program in biology. Walcott agreed. "But then," he warned, "you'll be just like the rest of us." As it stood, he told Payne, the entire program "benefit[s] from your way of being."[76]

It was then Payne realized that in every collaborative team of which she was a part—teams studying both whales and elephants—she made "the initial impression that [was] followed up on."[77] Payne was the one who called humpback whales "composers," who found their songs repeating, who felt the resonance of elephants' infrasonic signals. Payne, "simply an observer," was the one exposing previously unknown elements of animal communication with few, if any, of their scientific tools. After her initial impressions, biologists would pore over spectrograms, applying analytic tools to evaluate her hunches, but Payne did the listening. What is clear, then, is that analysis starts not with established technologies, but with a whole-bodied observation, with a certain comportment that allows the scientist to avoid jumping to discursive, interpretative conclusions. If analytic tools limit what we know, we must start with the opposite—that expansive feeling that exposes us to what we do not yet know.

Comportment, or the personal bearing of one's body, serves a necessary role in punctive listening and accounts for how we may attend to nature's punctum.[78] Yet attending does not always lead to experiencing nature's punctum, as Payne's hours upon hours of listening attests. Many instances of listening never resulted in hearing the songs of the humpback whale. Bodily comportment, it seems, serves as a conduit for listening, for placing us in a state of attention, but cannot guarantee findings. And other times, when we have not intentionally comported ourselves, something in our natural environment reaches out, prompts us to sit up straighter, and grabs our attention. When writer Alexandra Horowitz inquires what, exactly, attention is, whether an ability, a tendency, or a skill, she invites us to consider how much we can control about what we see, hear, or notice. We can actively attend, actively comport our bodies to heighten certain senses, but we cannot control whether nature's punctum will meet us in our moments of attention. Attention and comportment are thus linked, both integral to punctive listening, and subsist together only through patience and humility.

In practice for Payne, her way of comporting involved pairing observation with respectful reserve, a reserve that allowed her to withhold, or at least suspend, discursive symbolizing. Reflecting on what Walcott admired about her "way of being," Payne offered this description of her comportment in the field: "I was out there noticing things and saying things the way a two-year-old does, and respecting them the way an adult does. So you see, I would notice something and then I would respect it, and then I would follow it up a little bit." The reason children are often more inclined to engage in this process is that they "don't hide what they're observing."[79] They do not question it, second-guess it, analyze it, extract it. They just store it. In this way, children adopt a bodily comportment that insists upon bearing witness, upon keeping the punctive moment at the fore of attention. Payne, too, respects how the punctum is received and stored via indelible traces through the body—that is why her memories of first hearing humpback whale song involve leaning back in her chair and extending her arms, why her face changes rather than her words.

Payne's comportment may have derived from maintaining the curiosity of her childhood, as she describes it, but for those who know her in Ithaca, they say it has more to do with her musician's ear. "Did you know she's making violins now?"—that's the question I met repeatedly when individuals in town asked whom I was there to see. In upstate New York, it seems, Payne is a musician first and a bioacoustics expert second. McQuay, also an Ithaca resident, has often teamed up with Payne for his various NPR and National Geographic podcasts, and he reports that Payne "depend[s] on her ears in a way that many other people [don't]."[80] Most individuals employ their auditory senses unconsciously, McQuay posits, but Payne, as a trained musician, directs her ears consciously. Her bodily comportment is directed not by hearing, but listening, whether she's in the field, in the city, in a neighbor's home, or anywhere really. Her body is relaxed and "quite still," McQuay remarked—those are the bodily marks you notice with good listeners and trained musicians.[81] To be clear, Payne is not hearing more than a casual listener because her training offers more sophisticated taste or a better ability to judge the beautiful; rather, Payne hears more because her training has allowed her to develop a practice of sustained, patient comportment in which she can direct attention entirely to what is before her. Payne is almost always ready to receive a punctive appeal.

If punctive listening requires a bodily comportment that allows us to attune to nonhuman individuals, then it requires an understanding of self in which hierarchy is absent. In a word, humility is necessary to do the work of punctive listening. Consider how, in Payne's praxis, there is no clear sense of where her "meaning" begins and the whales' "meaning" ends, whereas in traditional analyses of semiotic encounters, meaning is thought of as being produced and received. As Payne describes it now, she always made sure to work with biologists she could trust to see "what their technical and analytical powers could add to my impressionistic and somewhat sloppy beginnings."[82] What she calls "sloppy," we can recognize as maintaining the "impressionistic" comportment needed to truly listen, upholding the humble bearings of a simple observer, an empathetic witness.

The Scientist as Witness

The desire to stay at the punctive level of "impression" and avoid discursive symbolic interpretation, it seems, creates an opportunity to expand how we consider the role of the "scientific witness." Witnessing resists the static nature of observation and instead opens up the relational processes involved in rhetorically being with others in the open semiosis. Or, as philosopher Kelly Oliver puts it, witnessing moves from recognition, or the idea that something (some knowledge or truth) can be attained, to the affective realm of experience.[83] "Witnessing as ongoing address and response between earthlings and their environments," Oliver contends, "cannot be reduced to recognition, mutual or otherwise."[84] When one-to-one recognition, and thus representation, is not the goal, then "vivid" experience can be valued, and that experience can mean in its affective register, altering our relationship with discursive interpretation. Similarly, multispecies ethnographer Chie Sakakibara defines a witness as one who "participates in the world in its relational aspect," one who does not center themselves in the world's radical relationality, but one who "stands with" others.[85] If observation is, epistemologically, caught up in objective ways of knowing that center the human scientist, then "witnessing" signals a way of knowing that is affective and experiential, a way of listening that surpasses the limits of the ear, a seeing not rooted in the eye, a knowing close and entangled with all that is before "us."

Case in point: when Payne writes about her work, her language shifts from "discovery" to "revelation." Perhaps a true mark of this witnessing in practice is when researchers believe there is nothing to discover. Payne distinguishes between discovery and revelation like this: "If you're thinking that everything is for people, then people will make discoveries for each other. And I don't think that.... I think things are happening right under our nose, right here."[86] The point is never to "own" what is under our nose, or to know it only in order to articulate it as a "discovery"; the point is simply to bear witness. And when what's right under our nose does not fit and cannot be understood with the discursive symbols and methodologies our disciplines give us, then we *simply* listen.

Witnessing is traditionally considered either a visual or discursive act: witnesses make present through words what they saw or experienced. If the origins of scientific witnessing, as detailed by historians of science Steven Shapin and Simon Schaffer, were not only discursive but also collective—the idea being that multiple, credible witnesses were required to turn an observation into an established "fact"—then bearing witness was about making experience definitive so that it could be accepted and trusted by others. Bearing witness in the schema derived from Payne's punctive listening, by contrast, is less about defining what was experienced, but about making present experiences that invite the imaginative interplay of aesthesis. The role of this scientific witness is not to define discursively, but to embrace the inadequacies of representation and make way for more than discursive ways of knowing.

Payne thus models a different sort of witnessing for scientists—which we will continue to consider throughout this book—but perhaps the most immediate implication of her praxis is how she makes listening central to witnessing and presents listening as participatory, a practice that compels others to realize their entangled relationality with nonhuman kin and broader forms of semiosis in natural environments. That is, Payne provides others— "scientists" or not—with the opportunity to have experiential encounters of their own. If all human animals have access to the open semiosis, as Kohn contends, then listening to nature need not be limited to academic, scientific endeavors. Perhaps this is the reason that Payne positions herself less as an expert and more as a guide. Her responsibility, as she sees it, is only to listen, and to make way for others to listen if they are so inclined.[87] The role of the witness, it seems, is not to define experience, but to present others with the opportunity to practice listening.

Payne presents others with the opportunity to listen most explicitly through her conservation efforts. "I don't think that I can, by talking, convince anybody to be nice to whales and elephants," Payne estimates, "[because] writing only reaches a small crowd. They have to have the experience themselves."[88] And when they do have that experience, Payne trusts, the whales and elephants will speak on their own accord. Of course, Payne cannot bring everyone to the ocean or to the rain forests in Africa. And it is clear that she, too, wrestles with the complexity of her affective method that must be translated eventually into discursive symbols and scientific ways of knowing. More problematic still is that Payne cannot determine what others hear when they do listen. "You can't stop people from being themselves," Payne clarifies, "and they will address the problem they perceive in their own way."[89] In the realm of discursive symbols, negotiation is of primary importance and perceptions will vary. But Payne is not so much concerned with what happens there. If she can draw people to the semiotics of nature and let the whales present their own forms of meaning, then, she seems to insist, there can be no misunderstanding. "It's very important," Payne beckons, "to study birdsongs and to see the analogies and to look at the songs of the lemurs and the songs of the frogs and realize that singing is a shared experience among many species."[90] We human animals may struggle to save the whales through speech, but perhaps we can learn to get there through song.

As Payne recounted after returning from a recent recording expedition: "The success of the whale project [in the 1970s] didn't come from anything people did, except make the whales' voices available to the public. The whales saved themselves." Approaching the fifty-year anniversary of the release of *Songs of the Humpback Whale*, Payne asked, full of hope: "It could happen again, don't you think?"[91] Payne's question is fair, considering that Roger Payne's and her efforts helped lead to the moratorium on commercial whaling by the International Whaling Commission in 1986 that still remains in place today, helping to save whales from extinction. Yet whether the songs of the humpback whale can still cut through our current world is up for debate. There is arguably more noise now than there was then.

It would not be hyperbolic to say that the lives of humans and nonhuman animals are predicated on whether the songs of the natural world can grab our attention, can get us listening. But if listening *for* something is the default mode of engaging with our surroundings, then, as Payne shows, bearing witness and punctive listening may require a steward who will present signs, songs, and rumbles on their own terms, allow them to make their own case. Payne becomes such a steward, not by speaking *for* the whales, but by simple observation, by bearing witness, by listening, and by inviting others to listen for themselves. This form of bearing witness does not entail articulating the "correct" interpretation of what is seen—we know that witnessing cannot do that—but rather allowing the sounds and songs that emerge from natural environments to articulate meaning that may not map so easily onto human discursive interpretations, meaning in all its full, messy layers, meaning that taps into something broader and more indefinable. Something more akin to beauty.

Felt Objectivity

Science in the realm of beauty, I suspected at the onset of this project, would not be so concerned with objectivity. With punctive listening as her method of inquiry, surely Payne thought of "objective knowledge" as I did, as that Haraway-esque "view from nowhere."[92] But there, too, Payne had much to teach: objectivity need not be so problematic. In fact, Payne believed that her definition of observation—not occurring until the observer is fully enmeshed in and listening to the open semiosis of the natural world—was closer to objectivity than my own. Objectivity becomes problematic when scientists manipulate findings *for* other people, when they are after discovery, rather than just seeing what happens. Should scientists loosen their allegiance to disciplinarity, to the need to make and convey only discursive meaning, then they may be able to fully receive what is before them. That, to Payne, is objectivity: seeing what is "just there ... what [has just] happened."[93] Further, Payne corrected, she has no "method of inquiry." Feeling, or her way of punctive listening, was not a tool, she insisted, but a comported way of being in the world. And rather than identifying where aesthetics ends and the sciences begin in her praxis or even how the two interanimate, Payne would articulate the entire process simply as "experience."[94]

"Experience" certainly fits with her expansive semiotics and stands to offer another feminist take on the remaking of objectivity. This objectivity begins with feeling, specifically with feeling life as it moves through us and around us. If Langer's "felt life" was feeling brought to life through artwork that allows us to know or assess our relationship with life itself, then Payne's practice is a sort of "felt objectivity," an objectivity that is led by feeling and simply observing the life before us, rather than explaining it away. A deep attunement with our own way of living may be the necessary first step in this felt objectivity, or as Payne remarks, "I just want to live, and find out what the world I live in is like."[95] A distance may still be inherent to Payne's objectivity, but this distance is not between subject and object. Instead, this distancing happens within the self when a greater swell of subjectivity keeps discursive interpretation at bay. In order to tap into the open semiosis, the observer must persist in the subjective accruing of affect that swells during punctive listening. This objectivity resists the human observer as a discoverer and considers their role as a respectful steward who bears witness to a broader set of relations.

Payne does not lessen her subjectivity in order to engage in punctive listening and access the subjectivity of the animals she studies. Instead, she insists upon an even more radical subjectivity, if not *subjectivities*—that is, attuning to subjectivities across species—that are informed by "experience" writ large. In this objectivity, dreams, meditation, and the beauty of music come to hold epistemic weight in scientific inquiry. Enter her home, and you get a sense for what living out "felt objectivity" on a day-to-day basis looks like. About an hour into our discussion, a chipmunk entered the conversation, scurrying underfoot from the next room over. Payne peered under her wood-burning furnace where the chipmunk seemed to have settled. Then she reached for the sliding porch door over my shoulder to open it a crack: "Maybe you'll want to go run around later," she said to the chipmunk, before sitting down and continuing her thought. Later, when she left the room for a moment, I noticed a "daddy longlegs" spider on its web occupying most of the square footage over a corner desk. There was no inside or outside for Payne, no space that belonged to some creatures and not others. All subjectivities coalesced in her home, informing what we may know and attend to that day.

Objectivity, in other words, does not participate in radical subjectivity if humans remain as the only voice in the work.[96] Some part of what we can objectively know must be felt, and the biologists in this project exemplify how to start as feeling observers rather than rational hypothesis-driven disciplinarians. The strength of Payne's punctive listening is precisely this felt objectivity, her challenging scientific methods for inquiry even as she distances herself from the sciences in a disciplinary sense. In the coming chapters, we will hear from others who have taken her invitation to listen more closely—as these listeners learn from Payne's way of being in the world, punctive listening as a method becomes more pronounced.

Because if there was one thing that changed the course of this study, it was how almost every biologist I spoke with, when they would ask who else I interviewed, had a story about Payne, how she encouraged their life's work, how she changed their way of being, how vividly she exists in their memories. For decades, her punctive listening has circulated—anecdotally—among biologists who study animal communication and behavior. Many name Payne as the most amazing person they have ever met, someone with a "kind of presence," someone who pays attention to you like no other, someone who makes room for others to listen alongside her. Others say Payne brings a certain "magic" to every room she enters. She's "open" and lets feeling and intuition direct not only her attention, but how she makes meaning. Payne, "underappreciated and underestimated for what she's done," has a way of making much of what we notice seem important.[97]

Even when I talk to strangers about this work, almost everyone seems to have a story about Payne's legacy, at least about what they remember thinking the first time they heard the songs of the humpback whale. Or rather, the stories they tell are about what the songs of the humpback whale made them do. For composer Annie Lewandowski, whale songs demanded to be placed on mixtapes in the 1990s, between songs of loss and love. For writer Ada Limón, the songs of the humpback whale inspired a poem, one about race and borders and the evolutionary flux that leads from sea to land, from unknown depths to constructed knowledge, and back again:

> Did you know that whales returned to the water?
>   It went like this: water, land, water. Like a waltz.
>
> I once had a record of whale sounds,
>   I swear I understood.
>
> It didn't matter what worlds they were under,
>   what language,
>   what depth of water divided,
>   the song went on and on.
>
> What I mean is: none of this is chaos.
> Immigration, cross the river, the blood of us.
>   It goes like this: water, land, water. Like a waltz.
>
> I am in no hurry to stop believing we are supposed
> to sway like this, that we too are immense and calling out.[98]

Would we still be writing poems about whale songs, I wonder, if we understood every rhyme and syllable? Or is it more important that we learn something about how to listen? Evolutionarily, it seems, our brains search for wonder. Thus, I am inclined to believe that it is this way with the open semiosis: beauty, in all its inarticulate wonder, is what breaks through, finds our depths, compels our listening.

3

Extractive Listening

>   Down in the sapphire ocean
>   the Humpback whales are singing,
> maybe about the wonders there,
> ....
>   ... even as high within
>   its green radiance of Guatemalan forest
>   a wine-throated
>   hummingbird's "sweetly varied outpouring
>   continues for the better part
>   of a minute"
>   —ah, if only
>   the whales and dolphins swam
>   in *that* green light, and heard
>   those tiny singers in their sea of leaves,
>   such arias they'd interchange
>   ....
>   of course
>   they sing together only
> in human words, never I guess in any
>   but English ones in fact—
>   ....
>   ... What are sounds,
>   and what are songs, that we can make them,
>   that we have ears to hear,
>   that on these tiny waves
> of air, of water, even of magnetism, we have made
>   the smaller ripples that we call Meaning
>   CARTER REVARD, "Songs of the Wine-Throated Hummingbird"

How songs come down—that's the overarching theme of Osage and Euro-American poet Carter Revard's collection containing "Songs of the Wine-Throated Hummingbird."[1] How songs pass from one generation to the next. How certain notes are picked up through cultural learning, but how the capacity for song itself is written deeply, genetically. Inherited, songs have much to say about what emerges in the world. It is one thing to ask, as Roger Payne did, whether song is thousands of times older than we are. But would it be too much to ask whether we are—all of it, nature, culture, the panoply of us—sung into existence? Because it's an evolutionary question, how songs leave throats with enough power to catch an ear, convince a mate to linger, and then get caught up in languages and lineages.

Although it may be a very human-centered concern to consider so much "song," to make sound mean in a manner that allows us to make sense of it. For Revard, this impulse—an ironic, vexed effort to translate between languages, mediums, and context—cannot go unquestioned.[2] As we learned from Susanne Langer, humans are drawn to the arts to make sense of our worlds, but an injustice may ensue when we take what we hear and turn it into what we already know. Reading Revard through Langer brings me to consider our desperation to make sounds into arias, duets, choruses, to articulate how we are so moved as a Western impulse. This impulse I consider shared by dominant science because, historically, scientists have not always been the best at listening.[3]

The history of scientists studying song in natural environments points toward an epistemic oversight: we study what we can see or hear. How often animal communication slips below the register of the human ear is largely unknown because until the 1980s, those signals were all but ignored. To understand what has kept dominant science from uncovering the full mosaic of sound that fills the airwaves, we must study what the human ear cannot hear. Sound that slips into the infrasonic category eludes human ears unaided by technology and challenges what this book has—until now—established as listening. How scientists pieced together that elephants, land relatives of whales that exemplify evolutionary convergence, also employ infrasonic rumbles that can be felt in the chest, but not heard by the ear, once again involves Katy Payne.[4] This time, Payne did not repeat recordings on surround sound. Instead, she sat in silence with elephants until this meditative state filled her dreams. For scientists to listen to what cannot be heard, they may have to keep dreaming—literally.[5]

This chapter outlines the work that led Payne to found the Elephant Listening Project at Cornell in 1999 as well as what elements of her legacy continue on in the project's work today. Increasingly more grounded in quantitative scientific practices, the project's lead researchers Peter Wrege and Daniela Hedwig grapple with the humble realities of studying a species on the verge of extinction. What some have called the most beautifully emotional creatures on Earth offer unique opportunities for considering what it means to listen and to expose how often dominant science listens hungrily, or extractively. If punctive listening was defined in the previous chapter as a method of empirical inquiry that dismisses discursive hypotheses and thoughts as long as possible to promote greater resonance and aesthetic connection with our natural environment, then this chapter considers what punctive listening is not.

Relying on the work of Indigenous sound studies scholar Dylan Robinson and Indigenous scientist Robin Wall Kimmerer, I define "extractive listening" as the dominant mode of listening in Western sciences. Punctive listening then serves as a way of listening otherwise, a way of listening that considers our relation to kin, those nonhuman relatives with whom we are always amidst, rather than listening only for interpretative value. Further, if one starting point for rhetorics of science begins when researchers are at the limits of language, then here we consider what resources biologists may turn to when they face those limits in the field. Listening to your subjects—getting to know them deeply, punctively—requires a degree of anthropomorphism, of centering our humanness in research enough to acknowledge our positionality in relation to the positionality of our subjects of study. Rather than the mark of impure science, these biologists invite us to see anthropomorphism as integral to felt objectivity and a sense of resonance with our natural environments.

The Feeling of Rumbles

Like whales, elephants have long evolutionary and social histories that remain all too mysterious for biologists, probably because elephants also do not fit neatly into laboratories. So, in 1984, when Payne learned that a baby elephant had been born at the Washington Park Zoo in Portland, Oregon, she asked permission to sit with the elephants for a week. Payne had "been wondering what kinds of sounds elephants make," and her only real goal at the onset was to "see what elephants are like," a goal she describes as "a very innocent, playful, childish thing."[6] As Payne sat with the elephants each day—for nearly every waking moment she was allowed in the zoo—she practiced what the previous chapter defined as punctive listening, that "simple" observation, the assuming of a bodily comportment that allows observers to "reground" in the open semiosis, to bear witness. Once again, Payne began to learn a method of communication other than her own. Payne describes this shift into a broader way of feeling and knowing as such:

> I noticed, little by little through the week, that I was feeling over and over again a throbbing in the air, change of pressure in my ears, that would occur when I was near the elephant cages, but not when I was in other parts of the zoo. And I knew just enough—perhaps because of the whale studies—to know that there is sound below the pitches of the sound that human beings can hear.

And lo and behold, we discovered there was a whole other communication system there that no one had known about. It was just below the frequencies our ears could hear.[7]

In October of that year, Payne returned to Washington Park Zoo with recording equipment and a few scientist friends to measure what she perceived. The team would confirm that what she felt as "throbbing in the air" were infrasonic rumbles sent as calls from one elephant to another.[8] Payne would spend the next decade recording free-ranging forest elephants across Kenya, Zimbabwe, and Namibia, and in 1999 she would create the Elephant Listening Project, a program focusing on forest elephants in the K. Lisa Yang Center for Conservation Bioacoustics at Cornell that is devoted to "conserv[ing] the tropical forests of Africa through acoustic monitoring, sound science, and education."[9]

The team's 1986 article about the infrasonic calls of Asian elephants—the species located at the Washington Park Zoo—signaled a paradigm shift in bioacoustics elephant research and elephant conservation more generally because Payne and her team uncovered that most elephant calls take place below what the human ear can hear.[10] Researchers were familiar with trumpeting, roaring, grunting, and ear flapping but had missed the rumbles. How elephants communicate culturally through their families and societies, calling one another as they do via infrasonic signals, remained a mystery. Most elephant calls are sent in the 14–24 hertz (Hz) range, Payne and her colleagues William Langbauer Jr. and Elizabeth M. Thomas reported, with the lowest calls coming in at 6 Hz; the lowest signal a human ear can hear is around 20 Hz. Calls last anywhere from ten to fifteen seconds and are traded back and forth as one elephant makes contact with another, each elephant rumbling through their own unique harmonic structures. Elephant conversations last an average of ten minutes, followed by periods of silence, from a few minutes to a few hours.[11] Payne was able to feel rumbles in the air because although these calls register below frequencies heard by the human ear, they are nevertheless forceful, with some calls measuring at an intensity of 117 decibels—as intense as rock concerts, which average 120 decibels.[12]

Field researchers had wondered how elephants worked together so synchronously for years. Ethologists, conservationists, and wildlife researchers Cynthia Moss and Joyce Poole—who famously pinpointed that elephants could be identified by the marks, scars, and shape of their ears—had long noted the "sudden coordinated movement of large groups of elephants" across great distances "when no signal was apparent to the observer."[13] Because infrasonic calls travel much farther than higher-frequency calls, their

FIGURE 4. Listening to elephants featuring Katy Payne. Photo courtesy of Daniela Hedwig.

occurrence helps to explain the sophisticated degree of social coordination among elephant populations. Two years later, Poole and Moss would join Payne to confirm that the same calls, frequencies, and force in Asian elephants could also be observed in African elephants. Their 1988 study built on the team's earlier study and confirmed that elephants use infrasonic calls to "coordinate some aspects of their reproductive and group behavior."[14] Four decades of research now mark ten octaves in elephants' vocal range and place calls into three distinct categories based on where the sound anatomically originates—larynx, trunk, and idiosyncratic—and identify a range of call types, such as rumbles, roars, revs, barks, cries, and snorts.

Whereas nearly all calls are social, perhaps the most remarkable is the "let's go" call. Best witnessed when a family is gathered at a watering hole, the "let's go" call comes after a matriarch removes herself from the scene, turns outward, and rumbles low and long until each member of the family is by her side and ready to march out together. "When, but not until, all members are gathered," Payne reports, "a final rumble is heard and the group moves away in unison."[15] Elephants' infrasonic calls travel up to six miles in open space because rumbles are not sensed only through the air, but also through the feet. Hypersensitive nerve endings in the toes, feet, and trunk, called Pacinian corpuscles, pick up vibrations through the ground.[16] Soundscapes of the African rain forest present human listeners with elephants' loud stomps, shrieks, and trumpets, but it is important to remember that what pierces our ear is the exception in elephant communication—not the rule. Elephants demonstrate a full-bodied listening that is sensitive to resonance, to feeling vibration.

Elephants live in small matrilineal family units of ten to twelve members in which one older female matriarch oversees younger females and their calves. Male elephants depart when they reach maturity and live independently, returning to their families only for mating. The matriarch, who can live upward of fifty years and is often several decades older than the younger members of the family, carries the wisdom her family needs to survive, wisdom such as where to find watering holes during times of drought and which ancient routes can be traveled to reach these watering holes some forty miles away.[17] This wisdom is also crowdsourced, as matriarchs share information with other close-knit families to ensure shared successes in foraging, hydrating, and avoiding predators. But, in an occurrence becoming more and more frequent, if a matriarch dies before younger females in the group come into their prime and inherit this shared memory, her family loses more than a loved one; they lose storehouses of learning, which elevates stress hormones among family members for up to fifteen years. Conveying the capacity of elephants to remember, writer Julian Hoffman remarks:

> They remember ancestral paths across plains that look trackless to us. They remember bevels in the earth that briefly brim. They remember seasons of hardship and seasons of prosperity. They remember each and every member of their herd, or even elephants they have not seen for many years. They remember where rain seeps away, where rain makes rivers. They remember human languages, ages, and genders, as revealed in a study in which elephants reacted defensively to the sound of adult, male Maasai speakers, who, as a people have been known to kill elephants, while ignoring the voices of young boys and women in the tribe, as well as those of the Kamba, a people who do not hunt elephants. And elephants remember the dead.[18]

Elephants, one of the only mammals on Earth to rival the lifespans of humans, have much to teach about the weight and role of memory—what is transferred, passed down.

Just how invested elephants are in the other members of their social group is evidenced by their greetings after long periods of absence from one another. Ask field researchers like Peter Wrege, who directed the Elephant Listening Project following Payne: these reunions are rowdy affairs, and how, exactly, bond groups know to meet at a certain location in a certain time after spending months apart, beyond distances at which calls can reach, remains unknown. Present speculation rests on circadian rhythms, but that hypothesis is not proven—the only thing known for sure is that elephants do not reconnect with loved ones by happenstance. Should you visit the Dzanga Bai at that time of year, Wrege explains what you would witness and how you might interpret it. "[You] see a female who suddenly stops what she's doing. She orients toward the forest and her ears are out and she's listening. And then her older sister walks in twenty minutes later." Wrege refuses to believe that this occurrence is coincidence. "I think that she *heard* her sister," Wrege remarks, and her sister rumbled, "'Yeah, I'm here.' And sister came." Of this much, Wrege is sure: these families "didn't necessarily meet in the forest three months ago and say, 'Hey, let's go to Dzanga next month.' . . . Then, again"—Wrege pauses—"they might." Wrege suspects that "[Payne] would say, 'Yes, they might.' . . . And they *might*," Wrege admits.[19] Payne would appreciate Wrege's healthy dose of doubt because, at first, before evidence could prove that elephants communicate infrasonically, before recordings were sped up to make low-frequency rumbles audible, all Payne had was a feeling in her chest—a sense that what she heard was significant.[20]

Shortly thereafter, a dream would heighten this sense of significance, clarifying that Payne's hypothesis may have emerged from her chest, her skin, her gut, but only because she attended to this punctive encounter. Payne realized through this dream that she had been invited into a trans-species conversation, one that came with a clear directive:

> I was lying in a deep, damp, warm grassy sward in the faint light of predawn. Close, in fact looming over me, was a swaying, silent circle of elephants. They were large and small, and several were reaching out with their trunks to sniff me where I lay, tiny and helpless. They swayed hugely, breathing over me for a long time, and then the largest female spoke in a voice that I heard the way you can sometimes hear in a dream, without vocal features, language, or sound. "We did not reveal this to you so you would tell other people."[21]

But Payne did go on to tell other people, writing a book about her experience and founding an institutional center devoted to elephant communication research. At least, things seem this way on the surface: Payne ignored this matriarch's directive. Payne herself does not weigh in to interpret the dream; instead, she interprets the directive as "an enigma . . . a mystery." Payne is certain of only this much: "If there was a kind of conscious connection between whoever that matriarch was . . . and myself, then she knew that I wanted to be one of them. I wanted to be part of the society, and that's why they revealed the way [their] society is working to me."[22] In this sense, Payne was not a human to the matriarch; she was something less species-specific, more like a fellow witness to natural forces.

EXTRACTIVE LISTENING 65

Payne's entanglement in this scene—communicating with the matriarch, discussing what it means to witness this scene—gets at a difficulty with witnessing trans-species encounters. No neat divide between where her senses end and wider aesthesis begins, Payne is caught at an interpretative impasse, one many scientists never face. For rhetoric scholar Casey Boyle, this crux raises the question of "how [we are] to witness what we are always amidst, never beyond." Because natural environments are multimodal, constitutive of experiences across a spectrum of sensation, witnessing must also be shared, dispersed, multi-perspectival, multi-temporal, incorporating imaginaries more than, not beyond, the human.[23] Whether witnessing can be shared in scientific endeavors in ways that are not extractive, especially when witnessing moves to interpretation or testimony, is more complicated still, given dominant science's history.

What if, for the elephants in Payne's dream, people posed the same trouble as for Revard's hummingbird and humpback whale? The matriarch's directive not to "tell other *people*," then, is key. The telling was not the problem—the people are. That is, how people make sound into what they want it to be, how people interpret meaning and package that meaning into something that can be appropriated, how people, especially people in Western cultures, have a tendency to extract, colonize, and make things mean "in human words, never I guess in any / but English ones in fact."[24] This way of people is not Payne's way, to be sure. Payne listens and attends to what is before her without integrating it into an established scheme; she listens "without vocal features, language, or sound." Her listening is not a listening *for* something, but a listening that is deep, simple, and punctive. That's why the elephants are not worried about her. The elephants are worried about those who listen to extract, about those who refuse to bear witness.

Hungry, Extractive Listening

Witnessing has much to do with listening, or at least it should. As Indigenous sound studies scholar and xwélmexw artist Dylan Robinson explains, the Halq'eméylem word "listen," xwlálám, means "to witness *and* to listen."[25] Another form of the word, xwlálámchexw, means "you are called to witness." Rather than "grounded in the visual," this witnessing qua listening has more to do with attention, or even how an individual positions themselves to attend to the subjects before them and, most importantly, how a witness acknowledges their positionality as a listener. In *Hungry Listening*, Robinson exposes the whiteness of sound studies and introduces anti-colonial listening practices that work in opposition to forms of listening that are extractive. Extractive listening is the dominant mode of listening in Western cultures, as listening, in the context of the classical music that Robinson examines, has come to be defined by whiteness and ways of perceiving shaped by settler colonialism. Extractive listening is listening without witnessing, without bearing responsibility to and being in relation with all we are amidst.

Defining extractive listening as "hungry listening," Robinson pinpoints hungry listening in how Western ears interpret art music from Indigenous communities, which is often marked by steady, rhythmic percussion building in intensity and vocalizations that mimic nature sounds, as through the guttural timbres of throat singing. The notion that Western ears wish to "interpret" these songs is key. Through formal and informal education systems, many Western ears are "'civilized' into 'higher listening' forms" in which listeners are "oriented toward recognition and identification." The ear is taught to listen for what can be identified, known, and integrated into individuals' existing ways of knowing. Hungry listening "is hungry for the felt confirmations of square pegs in square holes, for the satisfactory fit as sound knowledge slides into its appropriate place," Robinson argues.[26] Songs in Canadian First Nations communities, however, defy such quick identification, sometimes working in "more than representational" modes, not simply locating places or identifying actions, but serving as "the law of Indian people," as living, legal documents.[27] In other cases, as in throat singing, listeners are met with sharp, strong sounds more than the tender melodies associated with "art" music. Instead of being open to how these works can serve as a connector to the rhythms and laws of natural environments, many Western listeners dismiss the material as empty and meaningless. Or, to put it another way, too often Western listeners dismiss this music *as* music, and thus colonize what can be considered art music.

Although the sciences are far from Robinson's focus, there is much to suggest that hungry, extractive listening is also the primary mode of listening in dominant scientific practices. After all, scientific ways of knowing have historically presumed access to Indigenous land for field research and continue to do so, pointing to the Western sciences' entanglement with settler colonialism. Colonialism is always connected to land relations, environmental geographer Max Liboiron explains, and "even well-intentioned science and activism [reproduce these] colonial relations" by treating colonialism as a historical wrong, one long past.[28] If colonialism is treated only as historical, Liboiron contends, as the occurrence of a dominant power that took control over an area and forced its language and cultural values onto the individuals living in that area, then how that history persists through present-day practices goes largely ignored. Environmental science is a prime example: its research is directed toward noble aims of alleviating the effects of climate change, and yet scientists, often white settler researchers, need test cases and control samples from the land. When these samples are taken matter-of-factly, as if the researcher has automatic access to the land, then science extracts and perpetuates colonialism.

If otherwise "good" scientific work often operates via settler colonialism, then hungry scientific praxis without these good intentions echoes from an even more troubling history. Western scientists once measured the skulls of Indigenous people to determine their capacity for thought, a measurement that Indigenous education scholar Linda Tuhiwai Smith says "offends our sense of who and what we are." She argues that "scientific research is implicated in the worst excesses of colonialism for many of the world's colonized people."[29] By presuming access to land and "giving themselves the power to define," many Western scientists enact colonial power and extract, forcefully and hungrily, from the natural world.[30] What Liboiron and Tuhiwai Smith make clear is that extractive ways of knowing are often embedded in scientists' processes of invention, as presuming access to land occurs at the beginning of research in the data-gathering phase. Further, whatever is foundational to invention imbues the entirety of the research process, such as integrating or interpreting what is found only according to Western sciences or failing to report back findings to local communities. In this way, collecting data to confirm a hasty hypothesis or asking research questions to align with grant opportunities thus comes to sound a lot like hungry, extractive listening.

Robinson offers further forms of intentional, if not intentionally anti-colonial, listening that may transform biologists' relationship with the diversity of agents in the natural environments in which they study. Xwélalà:m is not a listening *for*, nor primarily a listening *to* a particular subject, but a listening *through*, a listening that is always in relation to kin. For Kanien'kehá:ka peoples, because all listeners are grounded in some way—in a certain place and thus maintaining a certain relation to land—listening is "always a listening *through*, or in relation with land."[31] Rather than listening for "use-value or the drive to accumulate knowledge," xwélalà:m is "a practice of gathering that takes place in non-goal-oriented ways."[32] Attention and intention remain, but focus shifts from what is heard to *who* all is involved in the listening. No longer is the listener the "sole subject of listening"; instead, "the *relationship* between listener and listened-to" moves to the forefront of attention.[33] This emphasis on relations—on how the listener is connected to all the lives, human and nonhuman alike, involved in the listening process—is how listening becomes inextricable from witnessing. If witnessing involves, as anthropologist

Kim TallBear suggested, "standing with" those subjects we are experiencing, then listening is required to know those subjects, to hold them in our ways of relating according to their own terms.[34] In other words, witnessing is not just about remembering a past event—as it is often understood—but about a mode of sensing a scene, respectfully, simply, in relation.

At this point, it is important to acknowledge that settler scholars cannot access xwélalà:m in its fullness, as such listening extends from Indigenous epistemologies. We may learn from the idea of xwélalà:m and allow this mode of listening to revise our own colonial impulses to listen hungrily, but to appropriate xwélalà:m would be another form of extraction that ignores Robinson's warning. Such is why, at the end of *Hungry Listening*, Robinson asks settler scholars Deborah Wong and Ellen Waterman to grapple with xwélalà:m and their own habits of listening hungrily. Reflecting on the difficulty of listening non-extractively, of "still[ing] our learned behaviors of intellectual processing—analysis, reaction, interpretation—[to] just listen," Wong and Waterman eventually arrive at composer Pauline Oliveros's "deep listening" as a practice of anti-colonial listening.[35] Oliveros proposed "deep listening"—or listening "to heighten and expand consciousness of sound" by "connect[ing] to the whole of the environment and beyond"—as a method of not just hearing, but listening with attention and awareness.[36] Deep listening is thus akin to "punctive listening," the theory of listening proposed in the previous chapter through Payne's practice of "simple observation" in the field, but with an important addition. The practice of deep listening holds the expansion of sound as paramount. Punctive listening, on the other hand, holds the expansion of environment, or one's sense of environment, as paramount. That is, "environment" here includes all who are involved in sensing sound, with sound as the connective tissue that stitches subjects together within an environment. As Payne wrote of her dream, punctive listening is about listening "without vocal features, language, or sound," listening not just for the expansion of the sonic, but for the interplay of aesthesis therein.[37]

Shifting away from hungry, extractive listening, then—if it is caught up in how scientists listen to, not *through* or *with*, their environment—would require a reorientation of the senses, one that may be promoted not only by a shift in the body, but also in grammar. Further decentering Western colonial practices requires a shift from nouns to verbs, argues plant ecologist Robin Wall Kimmerer. Kimmerer aims to blend scientific and Indigenous knowledges, but because uses of the English language have a bias for nouns, for seeing things and objects rather than lives and relations, she suggests that "an ethical revolution may depend on a language revolution."[38] When nouns direct focus to objects, scientists study what is static, what can be placed in stop-motion, those objects that can be observed by their borders, where one thing ends and another begins. In contrast, "many Indigenous languages [highlight] the animacy and agency of other beings" by emphasizing verbs more than nouns.[39] Verbs show present processes in motion, Kimmerer explains, bringing relations—our twisting and folding into all with whom we share natural environments—into the forefront of attention.

As Robinson defines the relationship between "listener and listened-to" in "a Western sense orientation," the listener does "not feel the need to be responsible to sound as we would another life," resulting in a relationship in which extraction from the environment can occur without pause or a second thought.[40] Revard hints at the same critique in "Songs of the Wine-Throated Hummingbird" as he brings animals to sing together in English words via nouns, arias, duets. As whales and hummingbirds sing together, we listen to their songs as from performers on a stage. Spectating and observing a display that could not otherwise be witnessed, Revard allows us to extract from the environment, taking song that is not ours and integrating it into a scene with which we know and are familiar. Like Robinson, Revard may not have had the sciences squarely in mind with this poem, but the remarks nevertheless pertain: language transforms and with that transforming comes the risk of appropriation. To witness the songs and sounds of natural environments, then, researchers need not a stage but a field, or to enter the field as a place that interferes with our senses, in which there is no singular scene to be observed but an ongoing encounter in which our senses pick up only so much. There, "listening in wild places," Kimmerer explains, "we are audience to conversations in a language not our own." To convey what we hear from the field, from those wild places, cannot come through present iterations of science, precisely because its language objectifies, codifies, silences. "Something is missing" from the scientists' vocabulary, Kimmerer continues, "the same something that swells around you and in you when you listen to the world," suggesting that what science steals in words, it may amend by listening.[41]

The power in listening comes in part from shifting our grammar, but also from shifting our sense of time. Extractive listening is quick, hasty, and feverish. So quick, in fact, that extractive listeners are unaware of how immediately we interpret and use what we hear for gain. "To be starving," Robinson explains, "is to be overcome with hunger in such a way that one loses the sense of relationality and reflexivity in the drive to satisfy that hunger."[42] Anticolonial, punctive forms of listening, by contrast, must slow down. When listening becomes more verb than noun, it becomes less hungry—it attends to action that is present, it does not rush to explain or codify action. Indeed, verbness does not equal swiftness. When the gerund "listening" loses its noun-ness, listening becomes an act that is never really finished or resolved, but is always slowing, attending to what Robinson calls moments of "heightened presence," moments in which the content of what is before us compels us not just to consider our relations but that also encourages connection "to a sense of place that is not my own, and slows down the pace of listening."[43] In the context of rhetorical invention in the sciences, we could say that deep, punctive listening asks researchers to linger in invention long enough to know their imbrication in a natural environment and to account for as much. The sciences, of course, are not the only disciplines or institutions to default to extractive listening, and a valid concern is whether slow, punctive listening is any match for hungry, extractive industries.

In the Presence of Elephants

To learn how to slowly, punctively listen in a time of great urgency, I returned to Ithaca, New York, some four years after this project began, to learn about the impasse of the Elephant Listening Project from behavioral ecologist Peter Wrege, who inherited leadership of the organization following Payne's retirement. Global Conservation estimates that elephant population numbers from the 1800s—around twenty-six million—dwindled to just above one million by the 1970s, with the proclivity for ivory, reaching its fervor around the 1930s, as the greatest contributor to population decline.[44] A wave of protection in the form of a worldwide ivory ban in 1989 offered some relief, but by the time Payne started the Elephant Listening Project a decade later, the Convention on International Trade in Endangered Species was already allowing sanctioned sales of "existing" ivory.[45] Today the World Wildlife Fund estimates that around 415,000 African elephants remain, with the greatest threat to their longevity no longer poaching exclusively but decreases in habitat resulting from human agricultural expansion and the effects of climate change—with temperature shifts leading to habitat shifts.[46]

Before I can learn about the Elephant Listening Project's conservation work, however, I am instructed to "look for the geodome." That is how data analyst Liz Rowland offers directions to Wrege's home. On the back roads east of town, you will come to a stop sign where one road meets another; there, look for a large geodesic dome, a compilation of triangled tiles, just before the woods: Wrege should be in there. I am driving to interview Wrege only after meeting everyone else on his team. Two months away from retirement, Wrege is building a geodome in his backyard, on his own, from the ground up, precisely because, Rowland explains, "he's a workaholic."[47] In the fourteen years that Wrege has directed the Elephant Listening Project, Rowland recalls him taking only a handful of vacation days—Christmas, maybe Thanksgiving—resulting in heaps of leave time accruing over the years. So much so that Cornell has forced Wrege to take his leave before retiring. Unfamiliar with spare time, Wrege is directing his energy to the geodome, a hub for conservation itself, collecting water, capturing solar power.

Rowland was not wrong; the dome is unmistakable. As I approach it, I think of what I have heard about Wrege besides his workaholic tendencies: anecdotal remarks about how Wrege is the "complete opposite" of Payne—his old friend and neighbor. As Rowland tells me, Payne is "much more able or willing to conjecture. She'll allow herself to speculate and to empathize." Wrege, on the other hand, is more "data driven."[48] The contrast does not stop there. If Payne is interested in animal behavior, Wrege is interested in acoustic monitoring. If Payne is home preparing lunch to share with me after our interview, Wrege asks me to visit a humid geodome.

Once inside, I ask Wrege how it feels to close the chapter on his illustrious directorship at the Elephant Listening Project, expecting to hear how he and his team defined best practices in conservation efforts by using passive acoustic monitoring to measure population numbers and observe behavior, not just of elephants but those who endanger the elephants. Instead, Wrege reveals the question lingering on his mind recently: Can he convince hummingbirds to enter the geodome? In one corner, Wrege imagines, he will plant heliconia, bright red "lobster claws" with tubes of rich nectar to attract the high-pitched sopranos. Next to the waterfall, which will be just to our left, may be other tubular plants like bee balm alongside a leafy, tropical spread. Maybe then, with nectar to spare and open-door access, hummingbirds will come. A thriving community does not just happen in a geodome; this polyhedron is a collection of parts, each of which must be installed and tended to just so. Even then, Wrege cannot control the hummingbirds; he can only make way for good relations. Watching Wrege think aloud about every detail of the dome, I realize how much has rested on his shoulders for the last fourteen years. He exemplifies what it takes to conserve an endangered species living an ocean away: a relentless work ethic, an automated persistence, and a healthy bit of humility.

Wrege was not always this way—the elephants changed him. For most of his career, Wrege assisted in fieldwork for behavioral ecologists pursuing basic science questions about animal behavior. Animal behavior research runs the gamut from open-ended studies that seek to understand complex behavior and cognition—think Jane Goodall and Katy Payne—to more minute studies that focus on causal connections that explain how or why some behavior occurs. In the latter, those basic scientific questions are considered more "fundamental," and the animal being studied is less important than the hypothesis being tested. Or, as Wrege puts it, the early dilemma in basic science studies is always: "What species can best help us answer that question?"[49] When the question was, ironically, about helping behavior, Wrege studied white-fronted bee-eaters in Kenya. Small tan birds with green wings, blue-brown tails, and ruby throats like a hummingbird, white-fronted bee-eaters who are nonbreeding assist their breeding genetic relatives in nesting to ensure fledging success.[50] When the question was about polyandry, Wrege studied wattled jacanas in Panama. Wading birds with brown backs, white wings, black heads, and yellow beaks, female jacanas mate with multiple males, resulting in paternity with males other than a female jacana's social partner.[51] Over time, Wrege developed a reputation as an expert leading research teams in the field and eventually desired to pursue his own research questions.

So, in 2006, when Payne ran into Wrege just as the Elephant Listening Project was hiring for her position, she asked him to toss his name into the ring for consideration. But, Wrege jokes, this is no success story; he would hear a resounding "no" following his interview. The problem was that Wrege knew everyone on the search committee and "couldn't fake *anything*."[52] The interviewers knew Wrege's frustration with administrative work, and because the Elephant Listening Project required raising every penny needed for salary, research, equipment, and so forth, the position seemed best suited for someone else. However, when the newly appointed director took a different position some months later, the committee turned to Wrege. Everyone at the Cornell Lab of Ornithology knew that things would change when Payne stepped down as director of the Elephant Listening Project, but even Wrege, who had not yet met the elephants, could not have anticipated the drastic nature of those changes.

One of his first tasks on the job was to get into the field and observe the elephants up close. Coming face-to-face with African forest elephants was unlike any previous field experiences for Wrege. In other studies, monitoring his subjectivity was easy. "What's that object?" Wrege would ask. "What question am I trying to answer?" In Gabon, at the Dzanga Bai, such questions fell away. Not long after Wrege climbed up onto the observation tower, he made eye contact with one of the elephants and was stunned into stillness. "This animal is looking at me and contemplating me," Wrege recalls. "There is [something] so *clearly* there."[53] When I ask Wrege to articulate what, exactly, is there, only one word seems right: presence.

To say that "you're in the *presence* of elephants," for Wrege, is to acknowledge the emotional experience that is being seen and considered by one of the smartest, most complicated animals on the planet, humans included. Studies on elephant neurology find that pyramidal neurons—those neurons that contribute to cognition, recognition, and memory capacity—are larger in elephants and show capacity for greater neuronal connections than in humans.[54] For Wrege, being in the presence of elephants means being aware that the elephant is not just "looking at me" but also "contemplating me."[55] Philosophically, the idea of presence, at least as Eelco Runia defines the concept in relation to humans, is "the unrepresented way the past is present in the present." "Presence" is about "'being in touch'—either literally or figuratively—with people, things, events, and feelings that made you into the person that you are," that immaterial something always tied up in meaning making.[56] This idea of presence, of what becomes present in human awareness, is in part what Wrege describes. But he means more than merely standing before the elephants; he is conveying what it feels like to get caught up in a past or a history that is not our own. As Wrege contends with the elephant contemplating him, he must also contemplate what is present for the elephant.

Maybe that's why it feels like humans could swap stories with elephants—you share your past, and I'll share mine—if only we could speak the same language. For a while, the Elephant Listening Project attempted just that: to complete an "elephant dictionary" that decodes various calls and rumbles. The idea behind the dictionary was to figure out what the elephant contemplates, and this direction is where animal communication research should direct its energy, according to researchers like Joyce Poole. For over fifty years, Poole argues, animal communication has been "stalled at description," identifying the types, sounds, and frequencies of calls without connecting those calls to meaning. "Translation," as she sees it, "is where [the field] needs to go."[57] Wrege inherited this dictionary from Payne, but sitting there on the observation tower, looking that forest elephant in the eye, he shelved the dictionary project. "What's the point of figuring out the [elephant] dictionary," he asked himself, "if they're dead?"[58]

The Role of Listening in Conservation

Wrege answered his own question by arranging an acoustic monitoring grid of fifty different recording units across 1,250 square kilometers in the forests of Nouabalé-Ndoki National Park in the Republic of Congo. Acoustic monitoring allows researchers to collect sound data at various locations simultaneously and further offers a more unbiased method than direct observation.[59] That is, acoustic monitoring methods are often used to measure population size and sonic activity in species that are otherwise difficult to observe, as in the case of whales, elephants, and many nocturnal species. With African forest elephants that spend most of their time in the rain forest—rather than in open clearings where direct observation is available—acoustic monitoring remotely details how many elephants are located in recorded areas. Most bioacoustic research projects stop there—contributing findings on population status, population responsivity to anthropogenic noise, and so forth. Wrege, too, could have stopped there with monitoring, but he opted to use the monitoring grid for another purpose: to detect gunshots, determine where poachers are localized, and report that information to local authorities.

Few individuals, if any, have spent more time listening to elephant rumbles than Liz Rowland, the Elephant Listening Project's "detector" who wades through nearly a thousand detections by hand in half an hour's time. Rowland has repeated this work daily and weekly for the last twenty years. When I ask what, exactly, an elephant rumble looks like on a spectrogram, Rowland admits that the team is "forever tweaking our understanding of what might be an elephant and what might *not* be an elephant."[60] Some rumbles are unmistakable—there's a wavelike pattern on the screen similar to a whale call, but also similar to an insect fluttering its wings near a recorder. If wave patterns appear only on one recorder, then you know what you see is an insect. But if the wave pattern appears on data from multiple recorders, then you have a definitive rumble. Gunshots look different—shorter, with a wider range of decibels.

Listening to these sounds—identifying gunshots and elephant rumbles among the soundscape of the African rain forest—requires a trained ear, not a scientist's ear or a musician's ear, but the ear of a local who knows the sounds of this particular forest. To echo Robinson, the positionality of a listener *matters*. Wrege learned as much when he and Rowland sat down to analyze sound files from the acoustic grid by hand. Hand-browsing these sound files involves listening to audio while looking at a spectrogram of the files—rumbles and gunshots present unmistakable patterns. Or so the team at Cornell thought. Once, Wrege was convinced that he heard gunshots but a community team in northern Congo disagreed. When Wrege insisted, the team corrected him: "Listen carefully," they instructed. What Wrege had actually heard was "a monkey hitting the branch."[61] To date, that moment stands out as one of the most important lessons of Wrege's career: when conservation requires listening, listening must be done responsive to and inclusive of local communities.

Thereafter, the Elephant Listening Project set out to reshape their research team, with the aim of hiring and training a team of community scientists who could code acoustic data and act on the data in ways that respond to community concerns. "We need something that incorporates the *needs* of local people and [that] still serve[s] conservation," Wrege argues, "and [we] need to build the capacity of local people so they can do the work."[62] When local communities are involved in conservation efforts, not just nominally, but substantively and centrally, then the team members discussing with families and communities why this work matters are not settler scholars from Western academic institutions. Community involvement as central to conservation further asks Western listeners to be humble enough to admit what our ears cannot hear. That Western ears have difficulty distinguishing gunshots from other sounds in the African forest shows how aural knowledge is situated, how listening requires a training of the senses that intimately knows and respects the most proximate sounds.[63]

Because human ears and the technologies that aid them have their limits, conservation efforts may still not be as adept as methods elephants have evolved to warn one another about dangers like poachers. Immediately following a gunshot, African forest elephants in Amboseli National Park display two dominant responses. First, a flurry of warning rumbles: elephants far removed from the poaching incident leave the park just after gunshots are heard inside the park. Later, rumble activity drastically decreases for several hours following the gunshot. This reduction in rumbling has even been traced to a given period prior to gunshot fire, evidence that elephants sense the presence of the poachers before the acoustic monitoring grid can detect poaching activity.[64] What I hear in these descriptions of elephant warning systems is evidence that witnessing is not exclusive to the human: elephants bear witness to *us*.

Indeed, elephants have much to teach human scientists about how to listen, about how sounds, calls, and memories come down through deep evolutionary time, about how to live in better relation with the natural world.[65] When ecologist Carl Safina outlined what elephants can teach us about living in relation to our environment, he came to this conclusion: "The elephants' way is humbler than ours. They demand less of the world. They take less from the world. They live in better resonance with the rest of the world."[66] Resonance seems to encapsulate most of what we know about how African forest elephants communicate and sense their environments. The etymological roots of "resonance" relate to synchronous vibration, or how vibrations in the air reverberate in other objects, amplifying and extending sound.[67] Communicating as elephants do through vibrational rumbles, it is not too much to say that elephants live in literal resonance with the world.

Resonance is a primary element in listening, too. If the eye makes "evident," according to French philosopher Jean-Luc Nancy, then the ear makes "resonant."[68] Vision concretizes meaning. Sound expands it. Just as the punctum punctures through feeling, sound functions in the same way, exposing us to meaning that is more than representational, in excess of what can be discursively depicted, meaning that resonates rather than represents. As Nancy puts it: "To listen is to be straining toward a possible meaning, and consequently one that is not immediately accessible."[69] To say, then, as Safina does, that elephants live in "better resonance with the rest of the world" means that their mode of engagement with the world is primarily through listening, not just hearing, but lingering in sound, feeling with the body, straining to build relations across time and space.[70] The elephants' way of listening is a listening through the ground, through the body, through time. Put simply, elephants remind us that resonance stems from a listening that is grounded, from a comportment of one's body in relation to land and others. For elephants, listening is more in the feet than the ears, and for humans, too, ears may not be the best guide for living in resonance with our environment.

After all, deep, punctive listening is not just about aurality, but also about bearing witness, about considering relations, about kinning. The concept of "kin" defines how all beings—all elements, really—are connected, entangled, related. Environmental humanities scholar Deborah Bird Rose famously asked researchers to center the idea of kinship in their work so as to ask "how to work toward world making that enhances the lives of others."[71] The need to engage in world making—or constructing ways of doing and fashioning ways of knowing that work to join ideas like nature and culture that have too long been treated dichotomously—requires a focus on action, not just ideation, which is why Kimmerer would turn us away from the noun "kin" and toward the verb "kinning." It is not enough to know or think how we are kin; we must also practice making kin, respecting relations, paying homage to life beyond ourselves.

Listening promotes kinning in that listening requires, if not consciously brings our awareness to, relationality, or how we are deeply related to other beings around us. Sound lends itself to kinning because "the sonorous [is] tendentially methexic (that is, having to do with participation, sharing, or contagion)," according to Nancy.[72] The visual, by contrast, warrants a different kind of engagement, one that is not so participatory. In theaters of ancient Greece, methexis was contrasted with mimesis as two different modes of audience engagement. With mimesis, the audience watched as action unfolded through the performance of a singular actor onstage. With methexis, the audience participated in the happenings onstage, often through improvisation.[73]

EXTRACTIVE LISTENING   77

Mimesis thus maintains a distance between listener and listened to that methexis challenges—methexis implicates the listener into the scene so that the listener must act and is unable to merely view the action. To put it simply, sound invites participation, calls for a response.[74] Langer, however, would not so quickly separate the visual (associated with mimesis) from the aural (associated with methexis). For her, visual art, in contrast to discourse, also means affectively, or more than discursively. What is important here is that whether we are seeing or hearing, we are always listening—punctively, deeply—to our environment, engaging with our environment affectively and methexically. Because, according to Langer, "nature speaks to us, first of all, through our senses."[75]

Anthropomorphizing, a Way of Kinning

To prioritize kinning in prescientific invention, we must consider what practices promote punctive listening and other methods for living out deep relationality. To be fair, it may be easier to make kin, to find resonance and relations, when you study elephants, or as Wrege puts it, when you are "in the presence" of elephants. Famous as elephants are for challenging scientists' objectivity, Daniela Hedwig, behavioral and conversation biologist who is now the third director of the Elephant Listening Project following Payne and Wrege, admits just how *involved* human researchers become in their studies of elephants. Because elephants' social interactions are so complex, the problem is that there is too much data coming in for biologists to be objective about all of it. It does not help, Hedwig explains, that watching forest elephants interact in the middle of a field clearing is a bit like watching a "soap opera."[76] Sitting atop the observation tower, Hedwig must study each individual elephant to determine just how everyone is related—who is the family's matriarch, who is mother to which calf, who fathered the calves, and how long ago—and these connections must be made as the family drama unfolds. Of course, describing the scene as a soap opera is an anthropomorphic move, but one that Hedwig contends is integral to good behavioral science, otherwise scientists may never learn who these elephants are as individuals.

Only when Hedwig begins to recognize the elephants as individuals do certain patterns of their behavior emerge. When their relations to one another are still a puzzle, Hedwig brings a sketch pad to the observation tower to draw and sketch all day. Placing pencil to page allows her to attend to each individual closely, noticing the slope of their ears, the folds of their skin, the marks on their trunks. "What are the features that distinguish individuals?" Hedwig asks herself. Then, she will notice "who is always displacing each other, who is actually sharing the water hole, and so [I come to] understand the relationships between them." For Hedwig, the social complexity of elephants—the cultures they create as family units and which they navigate and respect as bond groups—requires putting objectivity in its place. Objectivity is important, Hedwig explains, but only at certain points in scientific practice, only once a valid hypothesis has been constructed that warrants investigation. "In order to study animal behavior," Hedwig argues, "you have to let in that intuition in the first place."[77] Detailing how she manages intuition in the field, Hedwig explains:

> A big part of our work is to be patient and just watch. It takes a lot of time and it's kind of frustrating because you [just] want to *collect your data* and all that, but it's very important to sit down and observe the animals and try to put yourself in their shoes. That's not how I collect my data, but that's how I formulate hypotheses. This intuitive part [is] where I really try to put myself in [place of] the animal and try to understand their reasons behind what they're doing—that is absolutely important.[78]

Rather than a deterrent to objectivity, Wrege concurs, anthropomorphism may be a necessary heuristic in the study of social animals. When researchers study animals who maintain long-term bonds and relationships, they must consider individuals' unique characteristics as well as how those characteristics manifest in different sets and settings.

In the study of African forest elephants, for example, many scientists study their communication exclusively in bais, or forest clearings. Clearings are ideal for objective study—scientists can easily observe the elephants without the obstruction of trees. And yet, because forest elephants only spent 5 percent of their time in clearings, these "objective" measurements cannot adequately detail the whole of forest elephants' behavior. "If you're *too* objective," Wrege posits, then "maybe you can't crack those nuts, because you're not looking in the right place." The right place to study forest elephants, after all, is in the forest, not in open clearings where everyone gathers in what Wrege describes as a "bar-like atmosphere," crowded together in a tight space, talking over one another, communicating amid chaos.[79] It takes a degree of anthropomorphizing—considering, at least for a time, this social happening in the elephants' cultures as you would if you were witnessing this event in human cultures—to realize that bai interactions cannot indicate everyday behavior for the elephants. The result of not anthropomorphizing in this case is significant: if you take forest elephant communication numbers from the bai and use those metrics to measure estimates of elephant population numbers in the forest, readings measure erroneously low.[80] The issue is not that elephants are not present in the forest but that they rumble less in the forest just as humans talk less at home than they do at bars.

Overviewing the history of anthropomorphism in the sciences, Safina traces the stigma surrounding anthropomorphism to Konrad Lorenz, the Austrian zoologist many call a "founding father of ethology," or the study of animal behavior.[81] As Lorenz set out to prove that studying animal behavior could be objective, he adopted and emphasized an approach in which researchers "describe just what you see." The result was that "description—and *only* description—became 'the' science of animal behavior."[82] To describe and not to account for the researchers' thoughts, feelings, or emotions required the banning of anthropomorphism. By Safina's estimation, though, banning anthropomorphism resulted in notions of human exceptionalism—if we describe animal behavior only through description, humans become the only cultural, social beings, or at least beings who think, feel, and reason separate from nonhuman animals.[83] Other science writers agree with Safina, such as Brooke Williams, who argues that "anthropomorphism is a product of anthropocentrism," which places humans "as the most important element of life." And for Williams, assigning cognitive exception to the human results in vast limits to knowledge. "Thinking may or may not be unique to humans," Williams claims, but "what does seem to be our specialty is the belief that thinking is the only trusted form of knowing." He asks, "What have we sacrificed by limiting our knowledge to what we can think?"[84] One sacrifice rests in how, historically, the sciences have not encouraged methods that lend themselves to kinning. Yet "when someone says you can't attribute human sensations to animals," Safina reminds, "they forget that human sensations *are* animal sensations. Inherited sensations, using inherited nervous systems."[85] Rather than preventing humans from understanding animals, human sensations and emotions—aesthesis in its fullness—may be the best modes we have for making kin with our nonhuman relatives.

Placing what Hedwig proposed—that scientists limit their hypotheses if they try to be too objective too early on in observation—into conversation with Williams and Safina, we might say that researchers must allow empathy, intuition, and healthy doses of anthropomorphism into the early processes of prescientific invention to avoid misunderstanding animal cognition. Recall that Langer allowed us to define "prescientific invention" as the stage before objectivity enters, the stage of imagination when scientists attune to "vivid values" emanating from the scene before them, the stage that lends itself to slowing down and listening. Hedwig is a practitioner of "prescientific invention," even if she does not call being led by empathy in the early stages of data gathering by that name. "If that part of the process falls away," Hedwig warns, scientists "are not going to understand what's happening, because we need that process of really trying to understand *what's going on in their heads*, just by watching, and [being] patient, and using our empathy."[86] And yet, in the next breath, Hedwig points out the importance of objectivity during prescientific invention in that objectivity reminds the researcher not to interfere in the animal's behavior. Watching, listening, imagining what is going on in the head of the elephant, but without interfering—in this sense, the scientist is in the scene methexically but never takes the lead role. Anthropomorphism becomes integral to a felt objectivity that allows the human scientist to observe their positionality in relation to the positionality of their subject of study and respect the need to listen to all that cannot be known through our situatedness.

Committed to listening for as long as it takes for elephants to present their own ways of communication, Hedwig is now returning to Payne's elephant dictionary project, which is designed to help us "understand how [the elephants] think and what they feel, why they say [some] things in a certain context, and how they do it."[87] Consider the work as creating subtitles to their soap opera. But the stakes of the dictionary go beyond understanding their culture or even improving conservation efforts; Hedwig believes that the dictionary may provide new insight into the origins of human language, as elephant vocalizations are similar to human vocalizations, a convergence mostly overlooked by dominant science. For example, humans must modulate our vocalizations to produce vowels. Similarly, elephants use their trunks to modulate the structure of their rumbles, which Hedwig interprets as a "proto-vowel."[88] If human language is an example of convergent evolution, then the elephant dictionary, by naming calls and the contexts from which those calls emerge, may allow us to better understand connections to our elephant kin. By Kimmerer's estimation, naming is how "we humans build relationship, not only with each other but with the living world."[89] The dictionary thus becomes a way to act from punctive listening, naming and thinking with the elephants.

What I wonder, but do not ask Hedwig, is whether another motive for resurrecting Payne's elephant dictionary is preservation, with curating what we must remember but may not be able to save. Conservation has protected African forest elephants from extinction, but to date there remains little evidence to suggest that humanity will drastically change our behavior in the ways needed to prevent climate change. So, should there ever come a time when elephants no longer rumble, the elephant dictionary would bear witness to all the ways they resonated with the world. Preserving elephant communication through the dictionary, then, is a way of gathering what comes from kinning, of gift giving to those who will follow us, of sharing the beauty that filled our lives for a time. In every act of preservation and conservation, there is a lesson in beauty, about what it is that steals our attention, makes us stop, linger, and listen.

Listening to Beauty in the Field

If I was studying beauty, however, Wrege asked, why the focus on elephants? Elephants, he regretted to inform me, are *not* beautiful. Elephants are "a Dr. Seuss animal," just "so weird-looking" with the floppy ears and flailing trunks.[90] Truly, elephants throw ocular-centric theories of beauty in our faces, their beauty having nothing to do with proportions, curves, or colors. The beauty of the elephant is of a different kind, one closer to the bonds of kinship. Greeting loved ones after a long hiatus, inviting acquaintances to an ancient watering hole, issuing warnings across rain forests—these tenets of the elephant puncture, catch us unaware, have almost nothing to do with what we would call beauty in the cultural valence of aesthetics, but everything to do with beauty in the sensory valence of aesthetics. Listening to that kind of beauty requires patience, intuition, even anthropomorphism. Listen extractively and we are almost certain to miss it.

Or if we do find it using dominant scientific practices, Wrege now realizes, then it would take much longer than had we listened punctively. If there is anything Wrege has learned over his many decades in the discipline, it is that it takes the sciences decades to catch up to intuition. For years, Wrege observed that Payne's approach to scientific inquiry was "not respected" because she did not "toe that purist line." There were many years that even Wrege himself was skeptical of some of Payne's convictions about elephant communication and sociality. But if "you finally work long enough," Wrege finds, then you uncover time and again that "by damn, *she's right.* [Payne] said this! Thirty years ago!" If more researchers took the time to immerse themselves in prescientific invention, to study, as Wrege remarks, like Payne does without "any *pretenses*," if more scientists were led by intuition, by listening to beauty, then scientific practice may just move more swiftly.[91] The paradox of punctive listening is that it may, in certain cases, offer answers more quickly than the hungry pace of dominant science.

Wrege suspects that the younger generation of biologists are more open to these intuitive methods than his generation. And in hindsight, if he were to make any changes, Wrege says, he would have done more teaching in the field. He is doing just that now, trying to pass down these practices of punctive listening to his grandnephew. Walking through the woods, he encourages him: "Let's just sit on this boulder. Let's just listen. Let's just see what we hear." Wrege tells me, "My life has been rich because of the beauty of nature," and he hopes to pay it forward.[92]

When I ask Wrege who taught him these lessons, it is fitting, I think, that he shares a story about a woman, his grandmother, who fills his earliest memories in nature, who was intent on sharing with him "all kinds of beauty and wonder." Outside of Philadelphia amid acres of wooded oak, birch, maple, and pine that surround Swarthmore, his grandmother would take a young Wrege walking, sharing the names, colors, and textures of plants, stopping to listen to the observations of songbirds high amid the tree canopy. One day, he remembers, only now that we are talking about it, Wrege and his grandmother made a deal. Wide-eyed at the start of the trailhead, Wrege turned to his grandmother and said, "I'm afraid to miss some of what's beautiful in the woods today." So, the two struck a deal: "You look up, and I'll look down. Then, if [we] see something, [we] can tell the other so neither of us misses anything."[93] Listening to beauty is never a solo endeavor. Beauty emerges from our tangle of relations, and as it punctures our attention, it compels our listening, asks us to linger in relationality, to get caught up with our grandparents, the songbirds in the trees, the elephants rumbling in some distant forest.

4

Emergent Listening

> As technologies for taking life from the ocean exceed our ability to protect it, *Siren* focuses also on marine entanglement, one of the leading causes of death in the endangered North Atlantic right whale population. According to research at the Center for Coastal Studies in Provincetown, Massachusetts, 70 to 85 percent of humpbacks and North Atlantic right whales in the Gulf of Maine show signs of entanglement scarring.... This is a time ripe with urgency. *Siren* invites you into the tangle [with us].
> SIREN—LISTENING TO ANOTHER SPECIES ON EARTH

> Exceptions might be taken to the name bestowed upon this [Killer] whale, on the ground of its indistinctness. For *we are all killers*, on land and on sea.
> HERMAN MELVILLE, *Moby-Dick* (emphasis added)

To pretend that we are not all killers is too convenient a fiction. That much I realize, standing beneath a ceiling-to-floor, sail-shaped tangle of marine netting, mast-less and bare, considering how whales encounter humans, or how humans encounter whales, or how humans mangle our web of relations into such tangles that we cut life short. Caught up in the Herbert F. Johnson Museum of Art, I may as well be submerged in dark oceanic depths, as darkness surrounds all who enter this exhibit. Hues of pink, purple, and blue light spill onto the netting from different directions in stops and starts, colors synced up with the sound humming in my chest as much as in my ears—the songs of the humpback whale, some fifty years after the Paynes' famous album release.

The whales sound different, though, more modern somehow. Maybe it's because different colors signify different singers, and I'm learning each whale's individual sound, each unique voice. Or maybe it's because these recordings are new, and I'm hearing how the songs of some of the oldest singers on the planet have evolved over the last half century. But I swear these songs sound more familiar. There, you hear what sounds like a motorcycle pop, pop, popping as it slows to a stop, then rumbles a gurgle as it picks up speed again. A few minutes later, a screech that sounds like metal sprockets grinding together. Are these whales playing with human sounds, or am I, once again, anthropomorphizing? I will not test this hypothesis, but what if whales are imitating boat motors? After all, whale song is culturally learned, and the sounds of the sea are filled with anthropogenic noise more now than ever before—ships, sonar, seismic surveys—with some estimates measuring

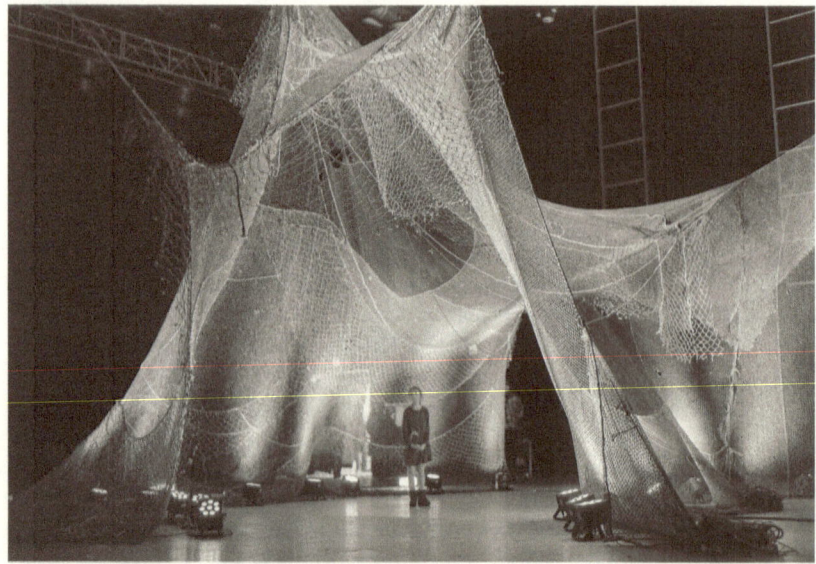

FIGURE 5. *Siren—Listening to Another Species on Earth* at MASS MoCA. Used with permission from Annie Lewandowski. Photo courtesy of Kyle McDonald, CC BY 2.0, https://creativecommons.org/licenses/by/2.0/legalcode.en.

a noise increase of over three decibels since the 1970s when the songs of the humpback whale were first recorded.

Or, as *Siren* asks me to consider: How much longer will whales be alive to imitate killers like us? What sound artist Annie Lewandowski understands as "unintentional whaling" inspired this exhibit.[1] Alongside artist and coder Kyle McDonald and scenic designer Amy Rubin, Lewandowski helped to create a scene that invites audiences to experience "the interior and exterior worlds of the humpback singer," their "creative intelligence" alongside the "dire circumstances threatening their survival."[2] Because humans are more paralyzed than motivated by fear, Lewandowski's team has opted to create a "call to action" through a "call to beauty." Without beauty, Lewandowski posits, people "can't muster the sustained response that crisis requires."[3] An enveloping experience that asks visitors to both marvel at and feel the weight of entanglement—in all its multitudinous forms—*Siren* composes a rendition of the songs of the humpback whale that earns "the ears of non-listeners," or challenges us to not just hear, but listen.[4]

As early as 2019, Katy Payne suggested that I speak with Lewandowski for just that reason: Lewandowski was an artist reaching audiences beyond the sciences, inviting those who had never heard the songs of the humpback whale to marvel at them for the first time. At that time, Lewandowski was listening to Payne's recordings of humpback whales, but she had not yet heard the songs of the humpback whale with her own ears, out on the ocean wallowing around in a boat, as Payne suggests everyone does, nor had she read *Moby-Dick*. But later that year, alongside singing humpback whales in the lee of Hawaii's Big Island, she was brought to the limits of her senses. Like Payne, Lewandowski is unsure how to talk about hearing the songs of the humpback whale out at sea. "I mean, I'm in my forties, and I think that I have an idea of what experiences will be like," she says, "but that experience taught me that I really have no idea whatsoever." She picked up *Moby-Dick* shortly after returning from Hawaii, and the line about how "we are all killers" has lingered with her ever since. Now Lewandowski is overwhelmed by the need "to be a better person, to tread *real* lightly on this planet."[5]

The previous chapter established punctive listening as an antidote to hungry, extractive listening. This chapter further defines punctive listening as a method of "emergent listening," or acknowledges the role of the human listener, that recognizes how witnessing adds different ways of sensing to that scene, and that acknowledges the estrangement—even pain—involved in comporting otherwise. Given present realities of anthropogenic harm, climate change, and mass extinctions, pain is inevitable, but as Indigenous ecologist Robin Wall Kimmerer has suggested, researchers must learn how to live, work, and act *through* grief and pain.[6] In the stressed state of our natural environments, the tendency is to linger on absence, on loss, but also on the idea that we humans can remove ourselves from such scenes. Here, rhetoric scholar Nathaniel Rivers points to our footprints: in environmental circles, those with the least impact, or the lowest carbon footprint, are considered the best kin.[7] Yet because impact is inevitable, because we humans are integral parts of our natural environments, having an impact is inevitable. Treading lightly or treading heavily, what is important is how we are comported as we tread, how we are attending and listening to those with whom we are in relation. Through the work of biologists Laura J. May-Collado and Michelle Fournet, who assess the impact of human sound on marine mammal communication, where sound carries for miles and animals are unable to escape the ocean's noisy chaos, we will consider how humans can interfere and disrupt the listening and communication practices of other species when we fail to consider how soundscapes—the sonic, acoustic elements that fill and make up environments—are shared.[8]

As Fournet and May-Collado parse through anthropogenic noise to study dolphin and whale communication, their approaches to listening in the field provide a more vivid understanding of the bodily comportment required to do the work of punctive listening. For Fournet, intuition cannot be forced, but her body can be placed in a way that allows for the kind of fully embodied listening that promotes intuitive leaps in inquiry. Rather than an ideal stance or position, this comportment estranges the body from its typical way of moving, sitting, and observing so that the scientist can know via feeling. For May-Collado, appreciation of beauty offers the estrangement needed to begin a line of inquiry, because it is there, in the midst of the unsayable, that something beyond what is already said, known, and felt can be sensed. These biologists, then, bring new valences of punctive listening into relief: listening to nature's punctum involves a sense of urgency, yet challenges how many people treat the emergency of the climate crisis. In what follows, we return to the waters and learn how listening emerges in our natural environments as well as how such listening may serve as a form of activism that invites others to feel the weight of the tangle of evolutionary emergence, if not better attend to beauty.

The Noisiest of Kin

The first sound that marine mammal evolutionary biologist Laura J. May-Collado remembers hearing as a child is noise. In the dark of night, the sounds of the Costa Rican banana plantation where she grew up blended into the noises spilling out of the rain forest, and May-Collado would lie awake, at times mesmerized, oftentimes terrified, and wonder about "how loud things were at night." That was the first biological question to ever linger for her: "What are the things making all these sounds?"[9] Listening to what she did not yet know, May-Collado defines her youth by the sounds of Costa Rica at night, mice singing and lizards clicking.[10] And she may very well have gone her whole life leaving the cacophony of the rain forest to her imagination—just listening in wonder, never deciphering—had she never moved to the city. There, in San José, frogs, crickets, and birds were replaced by motorcycles, buses, planes, and chattering strangers. There, May-Collado learned that noise is not created equal, and the experience was "a traumatic thing, especially at night."[11] When I ask what, exactly, changed when one kind of noise was replaced by another, May-Collado did not hesitate: beauty. For her, nature sounds may be the closest connection humans have to beauty, to feeling both the peace and pain of it. The problem is that in our attempts to amplify our own lived experiences, we drown out connections with nonhuman kin.

At least, sound is what brought her back to herself, to that young biologist lying awake wondering about nature's soundscapes. Pursuing a degree in biology in San José, May-Collado found herself without much sense of direction until a scuba diving trip off the northern coast of Costa Rica. Listening to dolphin sounds echo off the hull of the boat, she wondered, "What is it like to live underwater? Is it silent? [Or is] this space as noisy as the space that I knew as a kid?" Using the only affordable hydrophone equipment to which she had access in Costa Rica, May-Collado began to ask simple bioacoustic questions about the soundscape in which dolphins live, feed, and reproduce. She dreamed of performing the kinds of experiments she read about in academic journals and worried that she never would, as she noticed researchers in the United States who study bioacoustics "come with an engineering background, a physics background, experience in sailing, all these experiences that are not available to everybody."[12] Such is why, when renowned behavioral and physiological ecologist Douglas Wartzok invited May-Collado to study at Florida International University, she initially felt the need to ask more traditional questions about marine mammal "senses or behavioral observations."[13] But that work would not help her address the mystery that began in the Costa Rican rain forest. The noise May-Collado remembered felt like a puzzle, and her mind would not stop rearranging its pieces, because she was not only interested in the sounds and signals of individual species but also in how songs come down, inherited as they are through trans-species phylogenetic connections.

"Evolution was one part of the puzzle," she recalls, "but in the tree of life of whales, there were a lot of question marks" about the sounds dolphins and whales produce.[14] For example, what's the relationship between body size and tonal signals in whales? What whistle variations do bottlenose dolphins and freshwater river dolphins use to communicate? How has sociality influenced the evolution of sound complexity in dolphins?[15] But at the origin of every question for May-Collado is "how animals evolve in relation to sociality," which is another way of wondering how and why certain species communicate the way they do.[16] Take whistles in Ecuadorian freshwater river dolphins—biologists have considered this species as solitary animals, interacting not at all with other dolphins and yet somehow producing whistles, as do social bottlenose dolphins. The default in academic journals became to call the sounds of the river dolphins anything other than whistles, as whistles were established as signals that evolved for social purposes. Either river dolphins are not solitary or they are not whistling, the idea went. Because scientists knew river dolphins to be solitary, the dominant conclusion was that these sounds were not whistles. Such a conclusion, however, falls too neatly into dichotomous thinking, not allowing both things to be true: river dolphins are solitarily whistling.

Such misconceptions derive when biologists remain species-centric, rather than considering species' relation to one another. If researchers considered how river dolphins evolved from their ancestors, then they may be able to understand their behavior differently. So, May-Collado sketched a phylogenetic tree, or a tree that shows how multiple species are related, unearthing a puzzle millions of years old. When she placed the presence or absence of whistles atop each species in the phylogenetic tree, she uncovered that the whistles "evolve[d] earlier than the dolphins, at the branch that leads to all tooth whales." So maintaining a whistle or losing a whistle has nothing to do with sociality. According to her research, what *does* have to do with sociality is whistle complexity and variation: "the more social you are, the more complex [your] signals."[17] Bottlenose dolphins, for example, who live in family groups use whistles to identify one another, mimic one another, and work together cohesively. But their social worlds, we know, are not just made of species-specific connections. Dolphins live in marine environments overrun by perhaps the noisiest of kin: humans.

Now May-Collado studies noise almost exclusively, specifically how human noise alters dolphin and whale communication. After fifteen years of study, she can say this much: "The small resident population of bottlenose dolphins in Dolphin Bay, Bocas del Toro, is the target of the largest dolphin-watching industry in Panama," a reality that results in the species being at higher risk for extinction than other bottlenose dolphin populations.[18] Some dolphins in Dolphin Bay interact with up to a hundred boats per day—May-Collado and her team report that boat motor noise produces drastic whistle variation, an indicator of stress, and results in dolphins having to produce signals with acoustic traits that will allow them to communicate in a noisy space. In addition to exerting more energy, marine vessels separate individuals from their family groups and feeding grounds. The presence of vessels thus requires dolphins to travel more and forage less, an energy expenditure most pronounced for mother-calf pairs, whose energy stores are already vulnerable. Local regulations do specify that only two tour boats are allowed within a football field's distance (100 meters) of a dolphin in Dolphin Bay, but studies over the last decade have proven repeated lack of compliance with these regulations, with three or more boats within half that distance of dolphins nearly 70 percent of the time during working hours.[19] Careful study of changes in whistle modulation alongside analysis of the dolphins' habitat soundscape allow May-Collado and her team to determine how the increase of marine noise by 2.5–3.3 decibels over the last fifty years has altered the emotional state of dolphins.[20] Stress in dolphin populations is apparent by increases in signal complexity, with more loops in the signal and shifts in frequency, modulations that indicate sustained alertness.[21]

Before 2020, May-Collado and her team knew that vessel noise served as a major stressor for dolphins, but they could not say how much, precisely because there was no "control" in the natural environment, no marine area they could study where vessel noise was not present. No control, that is, until the COVID-19 pandemic created an unprecedently quiet environment in Dolphin Bay and across the ocean more generally. With no tourism boats circulating, shipping routes paused, and little recreational fishing, May-Collado and her team set out hydrophones to hear how dolphins would speak to one another when humans were not interfering so much. "The lockdown," they reported, "resulted in lower ambient noise levels and more frequent detections of dolphin sounds" along with "a more diverse whistle repertoire" in which signals "were longer in duration and less modulated than pre-lockdown."[22] Without human interference, then, dolphins communicate lower and longer.

Just as important as whistle modulation measurements are for May-Collado's work, considering dolphins' subjective experience in the milieu of their soundscape is equally important. Because, according to May-Collado, soundscapes are the only way to articulate a dolphin's "subjective story." We "cannot see expressions on their faces," but we do "know these animals rely on sound." We know that "their *world* depends on sound for them to find each other, to find food, to know where they are." Putting herself in the "context of the animal," considering the noise of their environment, May-Collado is brought back to her childhood in Costa Rica, to the noise that frightened and mesmerized her, but that nevertheless shaped her and that she desperately missed when she moved to the city. So, May-Collado wonders, "What happens if this world [of the dolphin] is suddenly masked with all these loud noises?"[23] What happens when snapping shrimp and whale song is replaced by boat motors and seismic signals?

May-Collado believes that if more humans asked that question and considered themselves as sharing a soundscape with dolphins and other kin, then we may just reconsider the privilege of noise making. Whenever she gives a research presentation about this subject, she asks audience members to talk with their neighbors and start up small conversations across the aisles. She then turns on a subtle boat motor through the room speakers. Slowly, she raises the volume louder and louder until audience members are yelling if not silenced completely by the noise. Her point is made: "Yeah, this is noisy. How do [dolphins] communicate like this?"[24] An anthropomorphic move that promotes kinning, or that allows listeners to consider their relations with marine mammals, May-Collado's tactic invites audience members to consider all the coping mechanisms humans have for noise—go to a quiet space, put on headphones—that elude our marine relatives. Immersed in a medium where sound travels over four times faster than in air, dolphins and whales have no easy methods of escape.[25]

When she is sitting at a computer, staring at frequency and signal modulation data, the beauty May-Collado sought to study when she became a biologist can seem far away. After all, she spends most of her time analyzing the noisiness, interference, and persistent disregard of humans. But this work is also precisely why science is more than a profession for May-Collado; science is a call to bear witness, to insist that humans recognize their effects and, more fundamentally, that humans learn the richness of phylogenetic trees, of genetic diversity, of how all the sanctity that must be protected is also a rich web of relations. Such a realization came to her through the words of mentor and behavioral ecologist Leticia Avilés: "Science may be a career for some, but for more of us it is also a calling, a calling to uncover the beauty that lies hidden in the remarkable biology of organisms."[26] Always, May-Collado's science begins with beauty. When she is in the field, asking her students to attend to beauty is easy. "We're seeing dolphins and whales, [and] we just take a moment to appreciate how gorgeous and remarkable they are."[27] More important than this moment of appreciation is the need for researchers to consider beauty when they are alone, charting phylogenetic trees or listening to soundscapes. In those moments, May-Collado realizes, beauty is not so close to the senses, but further up the phylogenetic tree, out there on the branches where aesthesis emerges from something earlier than any of us. Such emergence she considers as quite beautiful.

Emergent Listening

The time in which we are presently living, marked as it is by human-accelerated climate change, is one that many consider a "time of urgency." Recent studies indicate that Earth's surface temperatures have risen nearly two degrees Fahrenheit, with the ten warmest years all occurring since 2010 and the summer of 2023 measuring as the hottest on record.[28] Ocean temperatures are one to two degrees warmer than average, with a projected increase of an additional two degrees if drastic changes do not occur.[29] Warming temperatures result in melting glaciers and ice caps that have already increased sea-level rise by an average of six to eight inches, with an additional ten to twelve inches anticipated by 2050.[30] What these changes mean for humpback whales specifically is a decrease in population numbers by half, or at least such was the case from 2014 to 2016 when a drastic jump in ocean temperature, which marine biologist Adam Frankel calls the "Marine Heat Blob," covered areas of the Pacific Ocean on which migrating whales from the Gulf of Alaska are dependent for feeding as they make their way to the breeding grounds off the coast of Hawaii. This patch of warm ocean water, 2.5 degrees Fahrenheit warmer than average in some areas, began as an area roughly the size of the eastern portion of the United States—think, from the Mississippi River to the Atlantic coast—and expanded until the "Blob" covered nearly 2.5 million square miles in the ocean, from the Aleutian Islands of Alaska to the strip of Baja California off the coast of Mexico.[31] The extreme heat of the "Blob" led to plummeting phytoplankton numbers, which trickled up the food chain to drastic population decreases in cod, whales, and other marine life.[32] What this phenomenon means for marine biologists like Frankel who study how humpback whales interact and respond to their environment is that "there is no normal anymore" for researchers in the era of climate change. When there is no normal, research itself—its practices and ethics—must shift. With norms and controls no longer reliable, biologists must learn to better listen to what emerges from chaotic systems.

Temperature is not entirely to blame, but changes in temperature result in quick, unprecedented variation in sea level, sea-ice coverage, wave conditions, wind patterns, storm frequency, rainfall patterns, water acidity and salinity, and climate patterns more generally.[33] These shifting conditions thus change the ideal habitat where marine mammals live and feed. Whales depend on plankton and krill being located in a certain region at a certain time—when those places and times shift, whales experience nutritional stress and must increase physical exertion to locate available feeding regions, both of which result in decreased reproductive success.[34] Because the elements most directly impacting climate change are chemicals like carbon and human-made toxins like hydrochlorofluorocarbons (HCFCs) that waft into the air by the burning of fossil fuels and other industry production, reversal of these numbers or the prevention of additional damage is possible. Some researchers claim that if the burning of fossil fuels ended today, then air, water, and soil could capture and filter out all excess carbon in the atmosphere.[35] So, undoubtedly, the time to act to reduce emissions and protect life on Earth is now. Or the time to act was years ago, but the second-best time is now.

A sense of urgency, though, is too often linked with haste, that style of quick, hungry listening that leads agents to extract from their environments rather than remain responsive to it, listening carefully and punctively. Diversity and inclusion strategist Tema Okun has even called the "sense of urgency" a mark of white supremacy that results when individuals assume that the issues most pressing to them are most pressing to everyone, and it can be difficult to slow down and take the time needed to consider inclusive, democratic decision-making.[36] In the name of "expediency," too, existing power hierarchies are reinforced and hegemonic ways of knowing become the default mode of cognition, further marginalizing embodied or intuitive methods. Urgency, though, Okun admits, is the most difficult marker of white supremacy with which to contend, precisely because the push for racial, social, and environmental justice *is* urgent—the difference is where that sense of urgency is placed. White supremacy places urgency on short-term power, profit, and strategies, not justice.[37]

The push for justice requires a different sense of urgency, one that has less to do with hasty action and more to do with responsive ways of sensing and knowing, more with witnessing than detached objectivity. So contends rhetoric scholar Debra Hawhee, whose *A Sense of Urgency* is about feeling time, the synoptic magnitude of climate change and reckoning with the myriad time scales—the flux of evolutionary history and uncertain futurity—with which witnesses must contend. "The whole point of witnessing," for Hawhee, "is to bring the past to now," to right a wrong, to repair an injustice.[38] For without *in*justice, without the need for reckoning, there is no need for witnessing. According to public memory scholar Bradford Vivian, witnessing is about making present what *must* be remembered, which is another way of saying what has slipped out of the present but nevertheless must not be forgotten.[39] The sense of urgency that comes in witnessing is how the witness is responsible to what was witnessed, to what must be remembered. If witnessing has a motto, reports communication scholar John Durham Peters, it would be "we cannot say we do not know."[40] This scenario is exactly what separates witnessing from everyday acts of seeing: to witness is to be in the middle of emergency.

Consider the etymological root of "emergency." Talk of emergency brings to mind a frenzied state from unexpected, if not traumatic, stressors. But its original meaning, now rare, has to do with rising out of water, as in islands or icebergs emerging from the sea, or, more pointedly for our study, the breaching and crashing of humpback whales from fathomless depths—that's emergency. More generally, "emergency" has to do with the "process of issuing from concealment."[41] Rather than the idea of something hidden that comes into knowledge, the idea behind emergency is that something arises that could in no way be anticipated—what emerges is entirely outside the scope of expected possibility, in excess of what could even be imagined. That's why so many tragedies and emergencies interpolate witnesses—what must be remembered is what could not be fathomed, but that nevertheless occurred.

The connection of "emergency" with water provides a link to what we have learned about the connection between listening and being in relation to what we are listening to, in that the materiality of the term "emergency" evolves from the ebbs and flows of the sea. The root of "emergency" is "merge," from the Latin *mergere*, which means to "dive" or "plunge." And according to lexicographer John Ayto, the idea of a merger, or "combining into one," is a twentieth-century invention; before then, the idea involved that of "one thing 'sinking' into another and losing its identity."[42] To emerge, then, is to come forth *out* of a larger ocean, to rise into identity, with the unstated understanding that one may return to that greater whole. At its best, the scientific and theoretical concept of "emergence" still carries this valence of deep relationality amid distinctness because the idea of emergence formed to explain how things "arise from and depend on" but are nevertheless "autonomous from" one another.[43] In other words, emergence is "what parts of a system do together that they would not do alone."[44] Black feminist theorist adrienne maree brown extends this definition when she says that "emergence is the way complex systems and patterns arise out of a multiplicity of relatively simple interactions."[45]

Across these definitions, this much is clear: what is witnessed could not be anticipated and what emerges is in excess of what each individual brings to a given system or environment. Emergence, brown explains, "is beyond what the sum of its parts could even imagine."[46] Put simply, emergence is about relationships in a system, about how what comes from the water is part of and yet dissolvent into the greater sea. Borrowing from the "missing the forest for the trees" metaphor, we might say that some theories consider the ocean, other theories consider individual waves, but emergence seeks to define how the wave emerges out of the larger swell and is unable to be considered in isolation.[47] These examples of emergence surround us, brown teaches. Consider how birds migrate: they don't plan their trip, "raising resources to fund their way, packing for scarce times, mapping out their pit stops. They feel a call in their bodies and they must go, and they follow it, responding to each other, bringing their adaptations."[48] At the heart of emergence is the idea that, amidst change, something shared holds all the flux together, and how this flux unfolds may be witnessed in the minutiae of everyday occurrences, precisely because such minutiae are animated by phylogenetic histories eons old.

Further, as is the case with beauty, emergence is not only a phenomenon to be witnessed, but an ongoing practice, one that we are amidst, not studying from the outside. For brown, how we engage in processes of emergence are best depicted by writer Octavia Butler, who remarks: "Everything you touch you change, everything you change, changes you."[49] Rather than a causal chain of actions that can be traced to individuals, iterative change in emergent systems is about bearing responsibility for our place therein, understanding that we cannot avoid treading in our environments. Indeed, the very idea of emergence brings us to consider all the relations with whom we are entangled. The Proto-Indo-European "gere"—from which the Latin root of "emergence/emergency" and "urgency" derives—is older than our present formations, too, and carries the idea of "gathering" or "collecting."[50] Different from environing or surrounding, gathering has to do with "being gathered into," not doing the gathering. If environing is "being there," then gathering is an intentional practice of "being with" (or even a "living through"). We come forth *through* something, *within* something; we are never outside of it, we are methexically participating in it.

Emergence, then, must be witnessed punctively, always caught up as we are in the midst of it and unable to conjure or anticipate what may emerge. This synoptic nature of emergence is why many biological studies, as we have been considering here, resist the neatness of scientific method: hypothesis generation requires educated guesses, but emergence resists such guesses. To be honest about that, about the irreducible complexity that cannot always be brought before the human senses, would require punctive listening. Listening and waiting for what may never be able to be translated into neat words and symbols. In this way—punctive listening as a kind of emergent listening, or listening to what comes forth unexpectedly, more than discursively—would not be dissimilar, metaphorically, to echolocation. Dolphins are known to use echolocation in their marine environments, sending out high-pitched signals and listening for how the signal echoes in its return, echoes that reveal the size and distance of objects nearby.

When the biologists in this study employ punctive listening as a method, they also participate in the natural environment, comporting their body to send signals in such a way as to sense what information may or may not emerge from the field. Doing so acknowledges that the natural environment in which listening occurs is a shared environment and that listening requires an active, relational exchange. Emergent listening brings attention to what the act of witnessing adds to an environment: some meaning or relation is already there, but not previously accessible in a form that could be conveyed or communicated. Of course, as with echolocation, the act of witnessing is not promised. The energy we put into our natural environments may return unanswered, or if we sense a response, those signals may not return with information we are able to discursively convey. But meaning and communicating such meaning is not always the point with emergent, punctive listening. The point is to make kin, to be in a position to connect with the others with whom we share space, to acknowledge that the beauty of such encounters may always elude our definitions. Unable to train the eyes or ears for such encounters, the most researchers can do is prepare the body for listening.

Comportment, the Bodily Work of Listening

Not many people remember their first listening experiences—in remembering her own, May-Collado is unique. Fewer still remember how they were *taught* to listen, if they were taught explicitly at all. But for acoustic ecologist Michelle Fournet, remembering how she learned to listen makes her the bioacoustician that she is. Because before Fournet was a scientist, she was a singer who had, as she puts it, "the real privilege of being raised by big thinkers, of being asked to pay attention to the world around [her]." When she first sat down at a piano, Fournet recalls, her stepfather asked her to spend time holding down each note, her fingers pressing down, her back rounding toward the sound, her whole body listening until the note dissipated. "I'm *certain* that I would not be able to do what I do," Fournet insists, "had it not been for [that] training."[51] Her training in the arts taught Fournet that listening always happens in the body—you are not truly listening if you are not feeling what you hear.

This style of listening for Fournet lent itself to justice work, as exemplified and theorized by her stepfather, Ken Wilson, a philosopher whose "enclave theory" helped to make economic injustice in immigrant communities transparent. Fournet defines her stepfather's practice as one that encouraged her to practice "big listening," a listening beyond yourself, listening not just to music, but also "big listening to *people*." The idea was that "when people speak, they're saying things, and *how* they say it [matters], and that *language* matters." That's how, when Fournet began listening to whales, she knew immediately, intuitively: "Those aren't words."[52] Her training in music ensured that she could not extract or integrate what she heard into an established framework: she was taught to respect each sound on its own accord.

Fournet is adamant, then, that studying humpback whale communication must include more than simply studying their songs—our considerations of beauty when it comes to the humpback whale should be more capacious. After all, if only male humpback whales sing, then scientific study of humpback whale communication examines only half of the population. Missing almost entirely from the scientific annals are the characteristics, habits, and tendencies of female humpback whales, a Victorian-era carry-over that privileged males over females because of Romantic tendencies to locate beauty in song. Fournet considers this omission even more glaring because humpbacks are one of the most well-studied whales, probably *the* most well-studied whale on this planet, but until recently biologists were only studying male whales. To urge her field "to listen to the *entire population*" and privilege female voices, Fournet focuses her studies on calls, which are the foundational building blocks of whale communication and, yes, whale song.[53] By studying calls, Fournet is studying what is "innate" in whale communication, what is inherited, rather than learned. That is, song is "learned, cultural, built, cultivated, creative." But Fournet is, like May-Collado, interested in what "facilitates [that] social interaction," in how "what is shared" between individuals emerged in the first place. Even there, Fournet takes issue with saying that she studies "calls." As she puts it, "calls" are "not the most lovely way to describe *everything else* the humpback whales say."[54] And hearing the "everything else" is admittedly trickier than hearing song, which drone on for half-hour periods.[55] To catch calls in a repeatable way that can be studied and measured requires not just patience, but also a certain comportment of the body.

Although not the agenda of the documentary *Fathom*, an artifact of Fournet's listening remains in public record through the film, which follows Fournet and biologist Ellen Garland as they decode the calls of the humpback whale. Because of her conviction that the sciences should be as accessible as the arts, Fournet invited director Drew Xanthopoulos to follow her in the field, and the film makes clear the precarity of studying whale sounds—with one small boat not even half a whale's size, should these beings ever decide to make a toy of the marine vessel, they could.[56] Like the Paynes did, Fournet places hydrophones in the water, but her method to find the sound, to find where the whales are swimming that day, is less visual (imagine scanning eyes on the horizon) and more bodily. Midway through the film, we see Fournet hunched over in the boat, head resting on her knees, waiting to feel the sound of the humpback whale. She is waiting for the weight of their signal, the weight the note carries through the water, into her boat, through her feet, into her chest. When I ask about her technique in the field, Fournet explains that this position, this comportment of her body, facilitates what it takes to "experience the sound in the same way that [she] would experience a song or a symphony." Walking me through her listening process, Fournet thinks aloud:

> I close my eyes, often make myself small or close to the water. I do this when I have my headphones on too [listening to the hydrophone sound]. I'll physically *get smaller*. I think it helps me to *focus* on what I'm hearing. If I don't have to think about my body, if I am *set up* in such a way that my head gets heavy. Everything gets heavy. I rely on my bones to hold myself up, so nothing then becomes more of a distraction from what it is I hear.[57]

At this point, Fournet demonstrates, bending over in her chair, eyes closed, chest over thighs. The smaller she comports herself, the heavier her body becomes, or the more able she is to feel her body's weight. What Fournet does not say, but demonstrates, is that to know otherwise, the scientist must comport otherwise.

Only then, doubled over, feeling the weight of her body receiving sound, is Fournet able to listen "intuitively," a testament to how intuition requires bodily technique. Rather than merely a method for paying more attention, punctive, emergent listening allows Fournet to widen her field of perception, not via technologies but techniques. When she describes her practice of listening, she does so as if she's outlining a confession. "This is not going to sound very scientific," she says, "[but] I am an intuitive listener. I listen without agenda, and very, very *presently*."[58] What Fournet describes as listening intuitively is another example of what we have defined in this book as deep, punctive listening. Here, Fournet unearths how the body works as it listens deeply—the body serves, to borrow from Jean-Luc Nancy, as a "resonance chamber."[59] Fournet does have a list of "checks and balances," a "quantitative system that checks [her] intuition."[60] For example, to test the hypothesis of whether humpback whale calls persist across multiple generations in contrast to their rapidly changing song patterns, Fournet compared call types from 1979 to call types in 2012. Across the years, twelve of the sixteen different call types were substantially present in both recordings—scientific evidence of what Fournet felt intuitively: these sounds are innate.[61]

If emergent listening centers on the irreducible excess that comes from how beings relate in a given system or environment—about how organisms develop or how solitary river dolphins are able to whistle—then when it comes to whale songs, biologists who do not consider their simple, innate calls cannot fully consider how their complex songs emerge. Calls are the micro elements, the individual waves in the larger sea. Songs, Fournet has learned, are often prefaced by a call that Fournet and her team call a "whup," which they take to mean: "Hello, I am [so-and-so]." But even there, Fournet carefully tiptoes around defining this "whup" call as "translation," because, as she puts it, a "whup does not *mean anything*." Instead, "a whoop makes something happen. A whup *conveys* a state. A whup *conveys* identity." In other words, a whup is a conduit to meaning that cannot be discursively defined. As Fournet further clarifies: "[Because] how do you translate a sigh? How do you translate a sob? How do you translate laughter? All of those things tell us something about who you are and what you're doing and perhaps how old you are and make things occur, but there's not translation."[62] In this sense, Fournet seems to be describing whale affect, about what is conveyed in more than discursive modes. In doing so, she falls into the same slipperiness of meaning that rhetoricians have been grappling with for centuries.

Listening to nature's punctum, then, involves admitting what the scientist cannot know, to what is synoptically beyond language's grasp, to what we must comport our bodies into positions to feel, sense, discern. How what we witness emerges is a synoptic phenomenon that can never be presented through language neatly and discursively. We are, however, able to sense such emergence punctively. When nature's punctum affects us, we sense how what is before us emerged from phylogenetic processes beyond immediate comprehension. This is also how we are able to witness emergence, but not see it, exactly: we witness only what we could not have seen coming.

As Fournet listens to how whale calls emerge, she finds herself, like May-Collado, studying how whale calls, communications, and songs are altered by anthropogenic noise. In her own studies of the "Anthropause" that was COVID-19, Fournet and her team determined that as whale-watching trips decreased by nearly 70 percent, humpback whale calls in Skjálfandi Bay, Iceland, increased nearly twofold.[63] In the Arctic Ocean, the same was found in bearded seals. May-Collado and her team also confirmed this increase in calls for bottlenose dolphins off the coast of Costa Rica. These reports may even contextualize the occurrence of orca whales attacking boats off the coast of the Strait of Gibraltar beginning in 2020—perhaps these whales were responding to pain, to the reintroduction of noise that for a moment was silenced. Rather than wishing for animals to behave differently, we should pause and consider what these occurrences have to tell us about the effect of humans on nonhuman life. And at least in how Fournet and May-Collado teach, if these animals experience pain from our noise, then we humans should experience the same in kind.

The Pain of Listening

When May-Collado teaches classes on animal communication, her students have a choice to make: they will either engage in the process of messy, aesthetic "prescientific invention," or they will remain frustrated all semester. Or, maybe, they will experience both. Early on as a teacher, May-Collado noticed students entering her courses with the idea of science as static, established facts they hoped to memorize and regurgitate. What science really is, according to her, is an ever-shifting praxis that requires listening, aesthetically, to the agential subjects before you. Because human knowledge will always be imperfect, science itself is "always changing," scientists themselves "always learning." May-Collado admits, "That's why I love science, because it's not written in stone; it humbles you in many ways." That's the experience she wants her students to have: "They think that everything's perfect, [then] throughout the semester, they're frustrated, they're pulling their hair out."[64] Reaching the edge of knowing, the edge of sensing *should* be difficult and frustrating, if not painful—otherwise, scientists are just hungry listening, extracting data that can be interpolated into current existing frameworks for knowing. Truly listening to what emerges, to what could not be foreseen, is a slow, painful process.

At least, that's how May-Collado describes why she asks students to listen to hours and hours of field recordings as opposed to simply looking at and learning to analyze spectrograms. As she puts it, "The spectrogram needs to translate to something. [I] need to listen because it's important to associate that visual with the sound." Just like human communication, when animals communicate through sound, whether through calls or songs, May-Collado considers that they are, in a way, telling their own story. Take the songs of the humpback whale, which not only change over time, but change rapidly, with notes dropping in and out with each singing season. Listening to their song as you dissect each note on the spectrogram allows May-Collado and her students to "put ourselves in their world" and to consider "what does this variation mean?" Researchers may not know whether humpback song is a method for attracting females, competing with other whales, or vocalizations for the sake of play, but they "do know that this variation means *something*."[65] And it's important for May-Collado that her students not trivialize that meaning, that her students see these animals not just as research subjects—these are living, breathing beings who convey meaning in their own way, and scientists *can* get such meaning wrong. When she explains her rationale to students each semester, May-Collado says something like this:

> The only way to relate to [this animal's sound] is to listen to them so that when they're telling their story, you have that connection. So, [we] do not have that kind of dispassionate distance. [We] do love data, but [we] still need that connection because it's beautiful.[66]

Identifying and being honest about the beauty of these sounds is an important teaching moment: these things do not *have* to be beautiful, so if they *are* beautiful, we should attend to why that is the case. And we should recognize that those of us calling these things beautiful are also animals who emerged from the broader milieu of the environment we are studying and in which we are entangled. Our reaction to beauty emerges from something earlier, broader. Our responding to beauty sends its own message. And whereas we may not be able to know exactly why humans and other animals respond to and create beauty, we can say this much: beauty makes us pay attention and, if we let it, may just teach us how to listen.

Maybe that is why listening can be so painful, even if what you hear is beautiful: there are no shortcuts to listening. Even now, with sophisticated automated sound detectors, May-Collado forces her students "to actually go there and listen to all this data. Because if you *don't listen*," she teaches them, "how can you ask questions?"[67] May-Collado can count on students' initial response to her assignment: "Do I *have* to listen to all this?" When she responds in the affirmative, they ask, "How many days should I be listening?" or "Shouldn't we be creating an algorithm that finds the signal faster?" Yes, she replies, "we *could* be doing that. *Or* you could take the time to get familiar with this acoustic space. What does [this space] sound [like] at midnight? [at midday? in the morning?]" The process, she admits to them, will be painful and frustrating, but that's the point. "I like my students to go through the pain of listening—*a lot* of pain," May-Collado explains.[68] The idea is that, hopefully, as students persist through the pain, they will eventually get to the beauty, and—swallowed by beauty—they can begin scientific analysis that is connected with and in relation to their subjects of study.

When she and her students are enduring the pain of listening, they are in the throes of what Langer called "prescientific invention," where the point is not to integrate, solve, or hypothesize, but to listen. As they listen together, May-Collado tells her students, "I'm not using methods right now. I will use them later. But right now, I don't want to be practical. I don't want [you] to be practical. That will come."[69] In this prescientific listening stage, intuition takes over, or at least it should. And May-Collado's adage seems to be that if listening is not painful, then you may not be doing science responsibly, you may not be respecting all the rich cultures you are meeting.

If there is one thing that May-Collado wishes for her students to leave her class knowing, it is how not to bypass the listening. "I want my students to make the time to stop," May-Collado says, "stop that fast world where you are." You can always "go back to a book," she tells them, "but the tools [you] learn are very important in [your] future, and the tools of analyzing sounds can be applied to anything. But to learn, [you] need to listen." As they listen, May-Collado trusts, students will uncover the reason for listening, just as she did all those years ago, lying awake, listening to the rain forest moan. In truth, her experience listening to the rain forest at night is what May-Collado is trying to replicate for her students. "Stepping into the forest," she insists, imparts "a sense of urgency and a sense of awe and appreciation and *love* for nature." Sound, as she puts it, is "such an incredible product of evolution—random and just beautiful."[70] When you listen closely to beauty, in her experience, you might also find something that emerges, something punctive from within the recording.

Once, May-Collado and a student were listening to dolphins in Bocas del Toro and decided to place the hydrophone in a location where they did not usually record. Unable to sleep one night, May-Collado decided to listen to these new recordings. What she heard would keep her awake for many nights to come: a drumming sound that could not be assigned to the dolphins. Each night of the recording, with a time stamp of around 7:00–8:00 p.m.,

May-Collado heard a "Duh. Duh. . . . Duh. Duh. . . . Duh. Duh" sound that grew in volume as the night progressed until the drumming became "super loud."[71] After months of research, May-Collado learned that these sounds were coming from toadfish, rocky-looking bottom feeders known for their venomous nature. Listening more and more to toadfish, May-Collado and her student uncovered different variations of their sounds. Because researchers identify different species of toadfish living around Bocas del Toro, May-Collado and her student hypothesize that different sound variations may be reflected by different toadfish species. But first, before the hypothesis, there was just the pain of listening. Because of her patience, because she considers science as a humbling endeavor, "*because*," as she puts it, "I like to listen, I discover new sounds and then come up with different questions." No shortcut around the pain of listening, there are some sounds, in her two million minutes of recordings, that May-Collado still cannot identify. One sound, she describes, is "like a woodpecker" but underwater, and only at certain times of the year, and only at certain moments in those certain conditions.[72] May-Collado will continue sitting with the mystery of that sound and asking her students to do the same. The lesson, to borrow from Kimmerer, is that listening to our senses is a resource to science, not a detriment.[73]

Fournet describes yet another kind of pain associated with emergent listening—the pain that comes from connecting with kin in the field only to return to everyday ways of living that are totally removed from natural rhythms. Few researchers ask or talk about that painful side of beauty in the field, about how, when you get back, everyone treats the hours as "business as usual," how a deep sadness, a noisy cognitive dissonance, descends because you realize just how disconnected we are; how, in a sense, the entire world operates in what May-Collado calls "dispassionate distance." What no one tells you when you return from extended periods of fieldwork, Fournet explains, is that you will "not be understood when [you] come back."[74] Punctive listening changes you, and you may be unspeakably lonely when you leave a whole set of relations you have spent weeks and months learning. In the field, Fournet says, you learn to speak in different languages, process information in different ways of knowing, and "you can *devote yourself to something* and be with it fully, [you] can be present in that sort of mindfulness" that we usually only read about. Then "you have to take that experience and come back to a world that wants you to multitask, that is *completely* out of touch with the natural rhythms of humanity."[75]

If you survive the experience, and there is no guarantee that you will, especially when you study behemoths like the humpback whale, Fournet recommends connecting your future self with what will be your past self before you depart. For this experience, the scientist must be their own witness. Fournet's tactic is to write a letter to herself just before she leaves for the field and pin it to her refrigerator. When she returns, the letter is the first thing she reads, giving herself permission to remember, to bear witness to all she experienced. The letter, she narrates, goes something like this:

> If you're reading this, in some capacity, you have survived, and *no one* will understand what this moment is like for you. Writing this letter to you, I do not understand what this moment is like for you. You know things now that you didn't know before. Remember to read the [field notebook]. Go back and give yourself permission to experience it again. When you get lonely, you can relive it.[76]

"Remember to be kind to Luna," Fournet's small dog who stays home when she in the field. "[Luna's] in transition too," she writes.[77] When Fournet last returned from the Gulf of Alaska, she remembers coming home late at night, grabbing the letter from the fridge, crumpling to the floor by her front door, and sobbing, Luna clutched to her chest. A few hours later, she rose from the doorstep and went to work in a world that rewards quick, multitasking attention. Here Fournet narrates the dissonance between the aesthetic mode of punctive listening and the aesthetic mode of dominant science. While punctive listening, Fournet is a witness caught up and fully present with the humpback whales before her. When she returns to write up her findings, she feels asked "to be an independent, solitary thing," separate from the whales and pulled in disparate directions.[78] In a word, Fournet, like her subjects, struggles with all the noise of us humans, how our noise, too, often interferes rather than resonates with the soundscapes of the natural environments in which we are embedded.

The State of Attention

What has been implied, but not yet theorized, in our study of punctive listening is how these scientists are able to sustain the attention required to be fully present with their subjects of study for hours and years on end. Composer Annie Lewandowski, a former student of "deep listening" theorist Pauline Oliveros, offers a hypothesis: focus on the individual is paramount. These researchers must place the tendency to consider species behavior across a population in the background for a time so that they can learn to detect the unique intonations of individual beings, just as we intimately learn the voices of our loved ones from years and years of conversation. Case in point: Lewandowski spent over a year listening to three humpback whales in the creation of *Siren*. She learned their voices, their riffs, and their unique perspectives on song by giving them enough attention to consider what may emerge from their vocalizations. As I studied the exhibit over a three-day period, I began to worry that many of us will not lend the whales the same attention.

The composition of *Siren* considers the double nature of the term—its association with the ringing of warning or alarm became common only in the nineteenth century.[79] Originally, "siren" was used in the sense of "siren song," those singing creatures known through Greek mythology for luring sailors to their demise with the beauty of their song. When it comes to the songs of the humpback whale, both senses of the word are true: because warming ocean temperatures, anthropogenic noise, and marine entanglement threaten the future of humpback whales and other marine mammals, this time is "ripe with urgency." And yet, because their song fills the deep, this time is also one rich with beauty. Perhaps because of how beauty comes into relief when we realize we are losing it, visitors of the exhibit fall susceptible to the siren song of these largest mammals on Earth. Distracted by their beauty and longevity, we risk not allowing their songs to change our behavior.

But the above presumes that visitors will listen, will find these songs worth marveling over. Because I had spoken with Lewandowski months before *Siren*'s release at Martha's Vineyard—then later at Cornell as well as in Brooklyn, New York, and at MASS MoCA—I knew the musical piece to be forty minutes long. I also knew that Lewandowski had no plans to give explicit directions to visitors, no instructions that asked them to sit and listen for forty minutes as three distinct singers mixed their songs, each voice lit up by a different colored light. The placard that met some visitors—those who entered the exhibit through the main entrance rather than a side entrance—did offer those details of the composition, but for the most part, it would be up to each person how long they wished to listen. When the installation was at Cornell University's Herbert F. Johnson Museum of Art, those walking outside who were met with the wallowing bellows of the humpback whale could opt to walk a bit more slowly or sit on the lawn to experience these songs.

Lewandowski's past musical composition with whale song "Cetus: Life After Life" followed the same idea: each visitor could listen as they pleased, songs filling the atmosphere, crying out from the historic McGraw Tower, accompanied by hand-bell chimes, giving us the sense of what it might be like if we were bobbing in a boat and subtly began to hear the siren song of the humpback whale.[80] When I visited *Siren* at Cornell, whale songs echoed across the lawn of Libe Slope, growing louder the closer you walked to the museum. I sat on the green, anxious to hear how students considered the songs interrupting their studies. Most never mentioned these sounds. Some students would pass by the museum and yell, "Whales!" Others, more skeptical, wondered aloud, "What is this whale thing about, anyway?" My recording of the ambient atmosphere that day features whale moans beneath the backdrop of laughter, the chatter of concern about exams or plans for study sessions and weekend parties. Sitting there on the green, I wondered about the average attention span of museum visitors—would anyone give the whales five minutes, much less forty? So, I sat within and outside of the exhibit room for six hours over three days to determine how effective the whales were at convincing visitors to linger and listen.

Those who lingered in the exhibit joined me sitting on the wooden floor of the room, eyes searching the tangled netting, eventually coming to focus on the floor—stilled looking in order to listen. Hardly anyone took photographs, a testament, I think, to how *Siren* disrupted what visitors had come to expect from museums. No well-lit room, no small work of art that fits in a frame and lends itself to a quick photo—visitors are habituated to those rooms. A dark room with ceiling-to-floor nets and whale songs that bring a shudder to the chest does not call for the same behavior. I wondered how many knew that what we were hearing was whale song, how many had read the placard or known the sounds from previous experience. Sans photos of these marine mammals, how many could envision the humpback whale? But such is the experience of the ocean, I realized; you hear the whales before you see them, *if* you see them. Alone in this dark room, we were, at least sonically, experiencing the medium of water, not air.

No visitor stays the full forty minutes, at least not while I'm around to count. Even for me, whose job it is to sit and listen, I find it hard not to hop out of the experience and check my phone. Or when the yellow wood floors start to dig into my back body, I find myself drifting away to memories of high school basketball games, the squeal of the whale like the squeak of the sneaker. Remember those songs played during warm-ups? Something about nothing stopping us now. And eventually that first night, I realize that the time is late, no sunlight seeps in from around the corner, and no one shares the space. An attendant arrives, announces that the museum is soon to close, mentions that I am welcome to stay a bit longer, then lays beneath the arch of the netting. Both of us stare at the scene and the sound, and I realize that all day I have listened extractively, searching for something from the experience I can use, am using now, missing the point almost entirely.

Had I not been there for research, I likely would have interacted like the average visitor, listening for an average of two minutes and thirty-three seconds. Some visitors listened not at all, turning away from the room after only a brief few seconds. The longest anyone listened while I was present was fourteen minutes. What I see in many of these museum visitors, even in myself, is what Fournet describes as happening when she returns from the field—we all want to own a little of piece of "whaleness."

Whaleness

In practice, Fournet has experienced different impulses from individuals seeking her whaleness. At its best, wanting to own a piece of "whaleness" is an ethical impulse, Fournet believes, one not altogether misguided. The whale scientist must be "somehow better because [they] study whales"; many scientists are ethical, entering their profession to better human experiences and life writ large, but the whale scientists "must be *more* ethical."[81] Perhaps it is the biological impulse to nurture hidden deep within us—we envy, and wish to engage in, the whale scientist's conservation. Because it is a virtuous thing, is it not, to study whales in ways that promote saving them or, more generally, to make your life's work about communing with nature? Sometimes the impulse is more didactic, with people getting the idea that there is some profound experience or lesson to be learned by studying whales. Could we learn that lesson from someone else, perhaps we could bypass all the painful listening ourselves. Other times, the impulse is purely extractive. "How do I be you?" Fournet replays a conversation she has had too many times: "I want to go to Alaska. Take me to Alaska." What she hears in the dialogue sounds like "Give me this. Give me the whale. I want to steal your whaleness." "Have it! Take it!" Fournet wishes she could respond, perhaps flipping her long, thin blond ponytail over her shoulder as she does now in frustration as she relives those moments. "Have my whaleness. You can have the whaleness," she would say.[82] If "whaleness" is the state or quality of being a whale, then a person with "whaleness" has allowed the way of the whale to change their way of being in the world. People mistake "whaleness" as a thing to be had rather than a way of being. "Whaleness" cannot be had conceptually but must be practiced.

Inquiries for her whaleness "*cheapen* the experience," Fournet explains, turning the whales she has come to know as individuals into "things" to be owned, extracted, used. To appreciate whales as kin, Fournet opts to watch whales do what she considers "really boring things." She clarifies, "Like *really* boring things. I like watching whales breathe. That's my favorite thing that they do." Watching whales breathe, Fournet focuses on "how the whales perceive each other."[83] In a way, May-Collado wondered about the same when she asked of dolphins whether their soundscape was as noisy as the one of her childhood. Lewandowski, too, has a story about experiencing whales on their own terms, diving down as deep as she could in the vicinity of a singing whale to feel their song in her disoriented body rather than through her ears. Fournet's foray into that experience, though, was one morning when her vision was taken away.

Working on a whale-watching tour boat in Alaska, after leaving theater but before becoming a biologist, there was one morning where dense fog descended around the boat. This kind of fog—where you can barely see ten feet in front of your face—is tricky because if whale watchers were not able to see a whale, then each one was guaranteed a refund. With 140 passengers onboard, the crew felt the responsibility of that money-back guarantee. On this foggy day, the captain—a nineteen-year-old local whom Fournet calls "this really great, super badass lady captain who *very rarely* said anything at all"—turned off the boat motors and asked her to "go downstairs and tell stories." Fournet then spent ninety minutes of a three-hour tour downstairs telling ghost stories about shipwrecks and sailors being swallowed by whales. Suddenly, mid-story, the crew began running around, opening all the doors. Quiet descended, and "sure enough, you can hear the whales breathing." Fournet recalls, "You can't see anything, and then a whale comes up." The captain found the whales by sound. "What I didn't know then, and am only now *in this second* realizing," Fournet leans in to whisper, "is that might be the first time I experienced a whale the way that they experience each other. And it's extraordinary, and I *still do that*."[84] Fournet still places her hands over her eyes, lays down, low and small, and listens to what emerges, what could very well swallow us.

The mark of "good science" for Fournet is knowing "how [whales] experience each other."[85] Her human intuition is essential for "knowing how to interpret the natural world," but primarily because intuition guides scientists to the questions worth asking. According to Fournet, "teasing out [the] intuition" that punctively listens from the "ego" that extracts "is essential for finding truth, [and] truth is not just Western reductionist science," but also knowledges that emerge from land and seas.[86] Here, Fournet references Kimmerer's *Braiding Sweetgrass*. "I suppose that's the way we humans are, thinking too much and listening too little," Kimmerer writes. "Paying attention acknowledges that we have something to learn from intelligences other than our own. Listening, standing witness, creates an openness to the world in which the boundaries between us can dissolve in a raindrop."[87]

Nearly one hundred miles off the shores of Costa Rica, I learned about this dissolution of boundaries. I took Payne's advice to go to sea and experience what it meant to "wallow around in a boat." Along this coast, there was the hope but not the promise of whales, and with only clouds overhead, I stared into a horizon of onyx-black water. No whales emerged. Instead, I witnessed how the horizon ripples, how uneven peaks undulate, bowing and tearing at the seams. The hard, fast line of where one thing meets another is a delusion. The lines drawn—even neat, phylogenetic ones—between where "we" end and the evolution of us begins are not different. They rock, pulse, spill over into one another. There, I understand that beauty is not separate from those of us studying it, but an experience that emerges from the natural environments in which we are entangled. If Eduardo Kohn is right that "that which lives thinks" and John Durham Peters is right that "all life feels beauty," then perhaps one of the earliest ways to learn was by attending to beauty.[88] Which is another way of saying: Beauty is one of life's earliest teachers, instructing us how to pause, listen, connect, make kin.

5

Swallowed by Beauty

> The mower in the dew had loved [the flowers] thus,
> By leaving them to flourish, not for us,
>
> Nor yet to draw one thought of ours to him.
> But from sheer morning gladness at the brim.
>
> The butterfly and I had lit upon,
> Nevertheless, a message from the dawn,
>
> That made me hear the wakening birds around,
> And hear [the mower's] long scythe whispering to the ground,
>
> And feel a spirit kindred to my own;
> So that henceforth I worked no more alone;
>
> . . . .
>
> 'Men work together,' I told him from the heart,
> 'Whether they work together or apart.'
> ROBERT FROST, "The Tuft of Flowers"

Evolutionary biology is, in many respects, a study of lineages: how the webbing of feet echoes those of ancestors who navigated both land and sea; how aesthetic preferences shape the development of colorful plumage of blue, red, and green; how humans grew apart from other beings. These lineages hold echoes of the past and possibilities for the future. Such study, however, is often a blind endeavor, as there are only ever hints of how what we see now came to be. The lineage of how I found this poem—how the words of Frost came to me by way of ornithologist and evolutionary biologist Richard "Rick" Prum, and further how they came to Prum from Native American poet Carter Revard—reveals another shared goal, that of bringing humans into closer communion with nature. Following our interview, Prum asked about the other scientists and theorists I had engaged with in this project. It was then that I remarked upon the convergence between Langer's argument regarding biologists' "idols of the laboratory" and what Prum calls adaptationists' "physics envy." In response to this connection, Prum sent along "The

Tuft of Flowers": "[The poem] captures the feeling of discovering through your work that your concerns are shared by someone else you have never even known of—like Langer and yourself!"[1] And Prum had himself encountered the poem through collaborating with Revard. When the two discussed aesthetics and evolution in birds, nature, and art, Revard relayed the poem to exemplify, as Prum puts it, how "many separate lives [work] in parallel in different ways, in isolation and perhaps even ignorance of each other, toward a shared goal of discovery, beauty, and justice."[2] "The Tuft of Flowers" shows how unknown lineages inform present work, but Revard's own poetry brings into relief that it is not only we humans who work together toward these shared goals.

The butterfly, too, is part of this lineage, part of the web of relations that contains our differential becoming. Perhaps more than any other, that is the message echoed in Revard's poetry and emerging from this work on Payne, Prum, and others: human animals have much to learn about listening, about witnessing, about thinking beyond rational thought, and these lessons are perhaps best passed down by those nonhuman kin who came before. Kin need not be so species-centric, and whereas the other biologists in this book have focused their attention on mammals, Prum devotes his attention exclusively to birds. Revard, too, dedicates his most recent work, *How the Songs Come Down*, to

> the small birds only . . .
> who made their rainbow bodies long before
> we came to earth,
> who learning song and flight became
> beings for whom the infinite sky
> and trackless ocean are
> a path to spring:
> now they will sing and we
> are dancing with them, here.[3]

Like Melville's ruminations on the "antemosaic" existence of the whale, Revard is here humbled by the rainbow beauty of birds who flew over the land long before human feet trudged through its fields.

As with our previous studies, the case of Prum is an examination of a scientist who contends with the beauty of nature and who develops practices for listening to punctive impositions, or to thoughts, feelings, and symbols that mean beyond our present ability to discursively translate them. The difference is that Prum confronts this controversy head on: beauty, he argues, cannot only be proven scientifically, but has played a key role in evolution.

Through the work of Prum and his colleague Patricia Brennan, we learn how beauty not only has meaning but a tangible effect on the observable world, even if those effects are ineffable and better accessed through aesthetics and an open semiosis. This chapter presents Prum's theory of "aesthetic evolution" to reveal the centrality of a feminist epistemology of learning how to listen to beauty. Because Prum studies the colorful plumage of tropical birds, this chapter also considers what punctive listening looks like when a scientist's focus is primarily visual. In what follows, Prum and Brennan highlight not only how evolution protects the female's right to choose in avian and marine worlds, but also how the scientist must comport themselves in the field to witness nontraditional sightings. In the end, altering our relationship with aesthetics in biology allows us to better understand what interplay of aesthesis we may witness in our natural environments.

Rethinking Evolution, Aesthetically

If anyone delights in dancing with the "rainbow bodies" of small birds alongside Revard, it is Prum, avid bird watcher since age ten, winner of a 2009 MacArthur "Genius Grant" for his work on the morphology of feather development, and now head of his own laboratory at Yale University. Prum's established reputation in evolutionary biology stems from two groundbreaking findings: that feathers originated not in birds but in terrestrial dinosaurs and that the color of blue feathers is an example of structural color, or color that is produced from optical interference patterns.[4] Recently, however, Prum's work has taken a controversial turn in that he argues against biology's hegemonic interpretation of evolution through adaptation by natural selection and proposes instead an alternate hypothesis that he calls "aesthetic evolution."

Aesthetics in nature, albeit not biologists' primary focus, is not quite cause for controversy. After all, the beauty of birds' "rainbow bodies" is widely recognized and has been theorized as such at least since Darwin's time. The crux is that Prum has a different take on beauty and its meaning than adaptationists who equate beauty with fitness. As the running narrative goes, the greater a male bird's beauty (male birds primarily display more colorful plumage than females), the better his genes and thus chances for survival. The bright colors, the long tails, the ornate dances and postures speak to antibody resistance, strength, and vigor, adaptationists say. Alternatively, Prum argues that such beauty is almost entirely arbitrary, and has little or no relation to that individuals' fitness for survival.

At first, Prum's argument about the role of beauty in evolution derived from his subjective, lived experiences. For years he had watched the elaborate dances of the manakin and determined that fitness could not be what was on display.[5] Their moves—bobbing, weaving, and leaping over one another—were much too ridiculous, much too aesthetic. Yet saying that such ornament does not signal fitness is not to say that such beauty lacks meaning. According to Prum's "aesthetic evolution," the purpose is to appeal to the aesthetic tastes of the female. Prum thus concludes through a series of observations and other morphological studies that the beauty of birds is driven by female choice. As Prum attends to female choice, he radically shifts the foundation of evolutionary biology: if beauty no longer signifies fitness, if beauty is simply arbitrary, then biology seemingly loses its ability to predict evolutionary processes. Eventually, Prum would articulate the stakes for overturning the story of the adaptationists: a field dominated by masculinity and an almost exclusive view of how male animals control their environments could be overturned by making central the role that female agency plays in evolution. In other words, these dancing birds turned Prum into a feminist.

Before fully considering the argument between Prum and adaptationists, I should first define his theory of aesthetic evolution. It goes something like this: often in evolutionary biology only one side of sexual attraction is considered—the beauty of the dancing male bird. Aesthetic evolution contends that a full account of evolution requires an examination of both sides of sexual attraction, both the elements of desire (male coloration and ornamentation) and the form of desire (as it exists in the subjective tastes of the female).[6] Attuning to the form of desire reveals that many "evolutionary processes are driven by the sensory judgments and cognitive choices of individual organisms" and that most often in sexual selection, it is the female bird making this decision.[7] Aesthetic evolution thus speaks to the concerns of this project in two ways: first, considering female agency in evolution requires methodologies that are more subjective, more aesthetic, more anthropomorphic than traditional methods of scientific inquiry; second, disrupting hegemonic narratives of adaptation regarding male fitness requires feminist epistemologies that allow evolutionary biologists to see the female as a sexual *subject*, rather than a sexual *object*.

That's why Patricia "Patty" Brennan's work on duck genitalia is so important. Brennan is a biologist who completed her postdoctoral research through a National Science Foundation (NSF) grant under Prum's purview, and her studies reveal evidence that female ducks have evolved anatomical mechanisms to ensure autonomous mate choice in the face of forced copulation. Brennan's study extends Prum's aesthetic evolution by showing that sometimes female choice is enabled through aesthetics, sometimes it is ensured through morphology, but female choice is almost always ensured among nonhuman animals, a point rarely, if ever, acknowledged in evolutionary biology. All of which brings Prum to the question that now drives his research: "What does *she*, the individual female, want?"

To advocate for the individual within a discipline that desires generalizability, scientists must adopt a posthuman, feminist epistemology because they are the ones desiring generalization. Allowing the individual to speak for themselves requires a researcher who has no agenda concerning what is found and whether those findings fit with previously agreed-upon "laws." Turning his attention to the female—what does *she* want?—Prum considers the dance and wing songs of club-winged manakins across Central America.[8] These brown-and-black tropical birds may not have the coloration of the bird of paradise or the extensive moves of the Carola's parotia, but their wings create beauty of a different sort. To attract attention from females, male club-winged manakins vibrate their wings so quickly that the movement creates a percussive sound. Less a melody than a short, high-pitched beep reminiscent of a buzzer, these males have evolved to make such music. The proof comes in the morphology of their wings, claims biologist Kimberly Bostwick, another student of Prum's. Whereas most birds have hollow wing bones whose light weight promotes easy flight, club-winged manakins are the only known species of bird to have solid wing bones. These heavier bones that enable feather percussion come at a high fitness cost for males because they diminish flight capacity and create awkward maneuverability.[9] Prum explains that the club-winged manakins provide an example of "evolutionary decadence," or aesthetic evolution gone so far that it may bring these individuals to extinction.[10] That last point may be speculative, but how these solid wing bones challenge the adaptation paradigm is rather damning: if females selected mates based only on fitness, they would never select individuals whose bone structure impedes flight. Unless, perhaps, those females found the percussive sounds made by those bones rather beautiful.

Extending due credit to adaptationists, Prum does admit that ornate displays may sometimes signify genetic fitness, but such correlations between beauty and fitness he considers quite rare. So, while evolutionary biology's current null hypothesis—the standard that must be disproved in scientific experiments in order for data to argue a new point—is that beauty equals fitness, Prum proposes an alternative null, which he calls the "beauty happens" hypothesis. Should "beauty happens" become the new null, researchers examining any given sexual ornament or display behavior would have to ask: "Has the trait evolved because it provides honest information about good genes or direct benefits or because it is merely sexually attractive?"[11] Some of the time, beauty may speak to good genes—often, though, beauty is just beautiful.

Most of this argument is contained in Prum's *The Evolution of Beauty*, which reimagines how evolution works. And Prum's contemporaries do not just disagree with *Evolution*—they literally throw his book across the room. At least renowned evolutionary biologist Jerry Coyne did. In a blog post titled "Which Book Would You Hurl Across the Room?" Coyne selected *The Evolution of Beauty*. Coyne's anger reached its apex while "reading the bits about how human feminism is buttressed by observing the 'sexual autonomy' of female birds."[12] "It's wrong, it's smug, it's social-justice-y," Coyne continues, "and it's the prime example of the naturalistic fallacy."[13] Biologists Gerald Borgia and Gregory F. Ball also accuse Prum of committing naturalistic fallacies: "Prum's efforts to inject politics into science commonly distort the science to justify political goals. We should all have learned by now that science is about understanding what nature is, not what we want it to be. The arguments in [*Evolution*] that suggest otherwise should be rejected."[14] Put simply, a naturalistic fallacy states that because something is "natural," then it is "good," erroneously jumping from fact (identifying that something "is") to value (identifying that something "ought" to be).

To be fair, Prum never calls birds feminists—instead, he argues for aesthetic evolution as "a profoundly feminist scientific discovery." He clarifies, "It is not feminist, by accommodating the science to any contemporary political theory or ideology. Rather, it is a feminist discovery in that it demonstrates that sexual autonomy matters in nature."[15] An argument could be made, too, about glaring naturalistic fallacies latent in the entire field of evolutionary biology (e.g., male birds sing, so we ought to study male birds). Prum then argues that evolutionary mechanisms protect female choice, and that this finding ought to change how scientists conduct their research—namely, which "subjects" in nature they deem worthy of study. Critiquing what human sciences *ought* to be is not the same as making claims for what natural environments ought to be.[16]

Why Beauty Matters

The trouble with Prum is that at first it seems like he only wants to win an argument, specifically the argument about which Darwin is the right Darwin. An intricate theory of aesthetics fills Darwin's definitions of natural and sexual selection, Prum maintains, yet contemporary evolutionary biologists treat Darwinian aesthetics as "the crazy aunt in the attic that no one wants to deal with."[17] Talking aesthetics is considered "unscientific" by adaptationists. And to be fair, scientists have yet to design experiments that test for the subjective, arbitrary taste of nonhuman animals. Still other reviewers of *Evolution* convey the frustration of adaptationists: "By arguing that beauty is 'irrational,' 'unpredictable,' and simply 'happens,' [Prum] seems to simultaneously argue for the scientific study of beauty while setting it outside the bounds of scientific understanding."[18] As critiques flood in, even Prum is sometimes baffled by his dogged persistence.

The field of evolutionary biology has been critical of his work since Prum began building a theoretical foundation for the argument that "beauty happens" in 2010.[19] That beauty happens, that individuals respond to beauty for the sake of beauty, not because that beauty represents genetic fitness—these aspects of Prum's aesthetic evolution appeal less to scientists than to humanists like George Levine, who uses Prum to argue for a nature that operates under the tenets of art for art's sake. Levine's interpretation stems from Prum's use of "arbitrary" as a key term to explain how beauty comes to mean in aesthetic evolution—that is, beauty is arbitrary rather than a correlative to fitness. Prum defines "arbitrary" as such:

> Arbitrary does not mean accidental, random, or unexplainable; it means only that the display trait communicates no other information than its presence. It simply exists to be observed and evaluated. Arbitrary traits are neither honest nor dishonest, because they do not encode any information that can be lied about. They are merely attractive, or *merely beautiful*.[20]

Passages like this one bring Levine to declare scientific evidence for how "a poem, [or] a peacock's feather, should not *mean*, but be."[21] Because Prum researched aesthetics through the work of philosopher of art Arthur C. Danto, the idea of art for art's sake certainly persists in the present instantiation of his aesthetic evolution. As it stands, though, Levine takes Prum's argument to suggest that "beauty doesn't need meaning to exist."[22] Inherent in this use of the arbitrary, however, is the remnant of a more expansive meaning, an affective symbolicity that *does* mean, only not discursively.

Just ask the individual female what *she* wants. Researchers may not be able to know how or what beauty means to her now, but evidence suggests that beauty has certainly moved her and impacted the lineage of her offspring. As Bostwick remarks, historically what the female bird has enjoyed can be known because of those elements that are there before the eye now: "that big tail, the blue chest, the song."[23] Those features do not exist for the sake of themselves or only to call attention to themselves; they attract attention for the sake of selection: not only "enjoy," but "look at me, keep looking at me, select me." Beauty here does affect and tell of effects, but admittedly its language is not one that can be detailed through human logics. Therefore, the "arbitrary" speaks to an attunement with an open semiosis discovered through punctive listening. The "arbitrary" need not call scientists to become aesthetes (in the artistic sense) as much as it allows for more expansive meaning and the resources of perception in scientific encounters through alternative semiotics.

Even before the "beauty happens" argument, Prum almost always felt like an outsider in his home discipline and tried to come to terms with why that might be the case. While on sabbatical in Ecuador in 2007, he hoped to do just that. Prum examined the arc of his career and thought, "All [my work] is just weird [stuff] that Rick is into." Recalling that moment, Prum lowered his voice, leaned forward, and threw up his hands, "Which is working just fine. *Why complain?* I mean, I've got no reason to complain. I'm tenured at Yale, and I'm on sabbatical in the tropical forest with my children, and my wife is learning how to film hummingbirds. My life is not *missing anything*."[24]

Days later, sitting outside their A-frame house in the Ecuadorian forest, Prum would read Ron Amundson's *The Changing Role of the Embryo in Evolutionary Thought*, a largely unremarked-upon text arguing that scientists should attend to individual bodies if they are to understand how species change over the course of evolutionary history, an idea that will be more fully considered later in this chapter.[25] Prum remembers that he could not remain seated while reading Amundson's book. It was then Prum realized that here he was tenured at Yale but he was "[missing] a worldview," one that Amundson, as an evolutionary biologist, would define in part through attending to nonhuman individuals' subjective experiences. No longer was Prum alone; he had found the tuft of flowers left by Amundson. As Prum explains it: "I was the member of an intellectually lost tribe that had been *driven from the discipline*."[26] Through his study of birds, Prum thought, he might be able to emphasize Amundson's contribution. *The Evolution of Beauty* attempts such work by calling attention to the importance of individuals and, as an unexpected but necessary by-product, presenting a version of evolution not encased in "the masculinist, patriarchal structure of science culture."[27]

Before the emergence of evo-devo (evolutionary developmental biology) in the 1990s, few biologists considered that ontogenetic development "had *any relevance at all* to evolution."[28] Ontogenetics is the study of the individual organisms' development from fertilization to maturity (and sometimes for the entirety of its lifespan). Phylogenetics, on the other hand, is the study of the evolutionary history of organisms, not only the evolution of individual species but evolution within and among species.[29] Amundson—the scientist who Prum claims was "driven from the discipline"—sought to take down the adaptation approach to evolution through evo-devo's attention to ontogenetics by arguing that evolutionary biologists cannot understand the development of a species without first understanding the preferences of the individuals who embody that history. And although Amundson's argument was ignored, it would set the stage for Prum's aesthetic evolution and that theory's focus on the individual, especially the individual, autonomous choice of the female.[30]

A case in point for how often biologists overlook their individual subjects of study to generalize their findings and make their work seem more like "a real science," to use Prum's words, can be seen in the genre norms of academic articles in evolutionary biology. In one recent study, Prum reports that biologists studying begging calls wrote 2,200 words before they revealed what species of bird they examined for this particular study. By Prum's account, these biologists were "*ashamed of the subject of study.* . . . They were so eager to become generalizable" that they spoke exclusively of glucocorticoid responses as if "you could even talk about glucocorticoid physiology *as a thing* outside of any organism in which it could actually be manifested. It's certainly not true of [all birds], as if they are all the same."[31] In other words, in a desire to be more like physicists, more like those who are in the hard sciences, biologists have lost their focus on individual species, and thus even more so the individual. To uncover the evolutionary history of the present mosaic of natural phenomena, however, biologists have *always* had to approach individuals. The problem, Prum laments, is that the "whole field has been set up to try and hide that [focus on the individual]."[32] Aesthetic evolution attempts to return biologists' attention not only to the individual, but to nonhuman individuals' subjectivity.

One of the most important things to remember about individuals, Prum emphasizes, is that they do not obey laws. Individuals do not exactly break laws, but no neat system of laws can predictably be applied to them; their actions and desires are far too arbitrary. This idea seems radical, but consider how the current adaptation narrative of evolution forecloses the agency of individuals. Equating beauty with fitness brings scientists to believe that they can predictably determine what choices will be made in the natural world. To Prum, such a narrative leaves no room for individual subjectivity. Highlighting just how much of the natural world is filled with individuality, Prum explains:

> [Individuals are] not going to break the laws of physics. They're just no generalizations that apply to them because they are themselves unique. . . . One of the reasons why biology has so much vocabulary is . . . because we are exploding with individuality. *Genes* are individuals. . . . Homologues are individuals. These are historical entities.[33]

Allowing these individuals to tell their history requires acknowledging their subjectivity, their agency or subjective preferences, if scientists are to understand their becoming.

In this way, as the interview progressed, my view of Prum's persistence in wanting to win *the* argument shifted. Here was someone who had no reason to put his career and reputation on the line, but who did so because he feared that the wrong narrative of evolution could do harm, especially to women. Telling the story of when he finally decided to write *The Evolution of Beauty*, Prum relates a breakfast conversation with his youngest son. In 2011, Prum and Brennan had written an article on sexual conflict in waterfowl, which received incisive reviewer comments that Brennan calls the "worst of [her] life."[34] As a postdoctoral student, Brennan was concerned about how revising and publishing the article might negatively influence her career, so she asked Prum to shelve it. Prum, receiving her message one morning, was muttering to himself, frustrated, at his laptop when his son, then fifteen years old, asked an important question: "What's this really about?"[35] In response, Prum explained the details of duck genitalia: how male ducks have counterclockwise corkscrew-shaped penises nearly the length of their bodies, how forced copulation is common in duck sex, and how female ducks have, in response, evolved clockwise-spiraled channels in their vaginas to prevent unwanted sperm leading to paternity during forced copulation.[36] Here is why it matters: "*This is about coercion.* This is about sexual autonomy." Furthermore, Prum would tell his son:

> This matters to *you* because here you are a young person. One of the implications of this whole thing is that the real, healthy goal for young [adolescents] is to understand that the [idea] of sexual development is to become a sexual *subject*, not a sexual *object*, especially for young women. . . . This scientific problem we're having in this paper about ducks [is] connected to this huge problem in the culture, most of which was in and of itself propped up by other really bad examples of unconsidered patriarchy lurking, baked into our science, and [scientific] authority supporting all this stuff. . . . This *really* matters because young people like you ought to grow up in a world where they're not faced with this stuff, where they realize that science doesn't support this self-damaging view of the self.[37]

Afterward, his son asked how many people understood the stakes of this argument stemming from duck sex. Prum, realizing that the answer was far too few, wrote back to Brennan: "Don't let them knock us down. We are *on the right track.*"[38] The two would go on to revise and publish the article in 2012, and, on that same morning, Prum would drop off his son to school, return to the office, and start writing the chapter "Make Way for Duck Sex."[39]

The "Bizarre" Beauty of Nature

Duck sex may play a central role in Prum's theory of aesthetic evolution, but Brennan is the one who brought the realities of duck genitalia to his attention. Brennan had been studying the great tinamou—a tropical ground bird somewhat similar to turkeys—in her home country of Colombia when one moment during field research in Costa Rica changed her entire outlook. Before then, researchers had paid little attention to bird genitalia, and it was thought that birds copulate exclusively by touching their cloacal sacks together. That day, though, Brennan watched two tinamous mate, and their ritual struck her as odd: instead of a brief cloacal touch followed by separation, the tinamous stayed "attached" to one another and walked around together for some time. When they separated, Brennan saw "this little thing hanging off [the male's] cloaca." She jokingly remarks that before that moment in her career, "the words 'bird' and 'penis' had never come together in a sentence."[40] Because only 3 percent of birds have penises, witnessing a penis on the great tinamous was extremely rare.[41] But if some birds have penises, Brennan wondered, then evolutionary biologists know even less about sex than they think. Moreover, how did the other 97 percent of birds lose their penises? Brennan won an NSF grant by asking such a question, and shortly thereafter she entered Prum's office to inquire whether he would oversee the project. By the end of the meeting, Brennan recalls, Prum was "more excited about the project than [she] was."[42] Brennan's research would change what scientists know about sexual conflict in waterfowl, but Prum would see it as evidence supporting his alternative theory of evolution. Here, female agency is protected by the evolutionary morphological structure of a clockwise-spiraled vagina.

So unlike Prum, "beauty" is not a word that often factors into Brennan's corpus. Instead, Brennan uses a different term just as frequently: "bizarre." Over the course of our interview, Brennan referenced "the bizarre" eleven times, and the term is included often in her popular science writing. When I pointed out her reliance upon the term, Brennan threw back her head in laughter. She had never noticed the repetition, and she could not recount when or how the term became indicative of her observations of natural phenomena.

In lieu of an answer, Brennan gave me a bit of homework. Before I could type up our interview transcript or return to writing, I must watch *Dancing with the Birds*, a documentary that overviews the mating rituals of birds

SWALLOWED BY BEAUTY   119

in tropical rain forests across Papua New Guinea. Brennan had watched the documentary with her two daughters the previous weekend, and—she smiled while remembering—its contents left them in tears. Laughing so hard that she hunched over in her chair, Brennan blurted, "Oh my gosh, the males are ridiculous! They are *ridiculous*." She continued, "It's just phenomenal . . . because very quickly you get the whole picture of how ridiculous some of these [mating ornaments] really are. Completely arbitrary."[43]

FIGURE 6. Male king bird of paradise. © Kenji Aoki. Used with permission. Originally appeared in Ferris Jabr, "How Beauty Is Making Scientists Rethink Evolution," *The New York Times Magazine*, January 9, 2019.

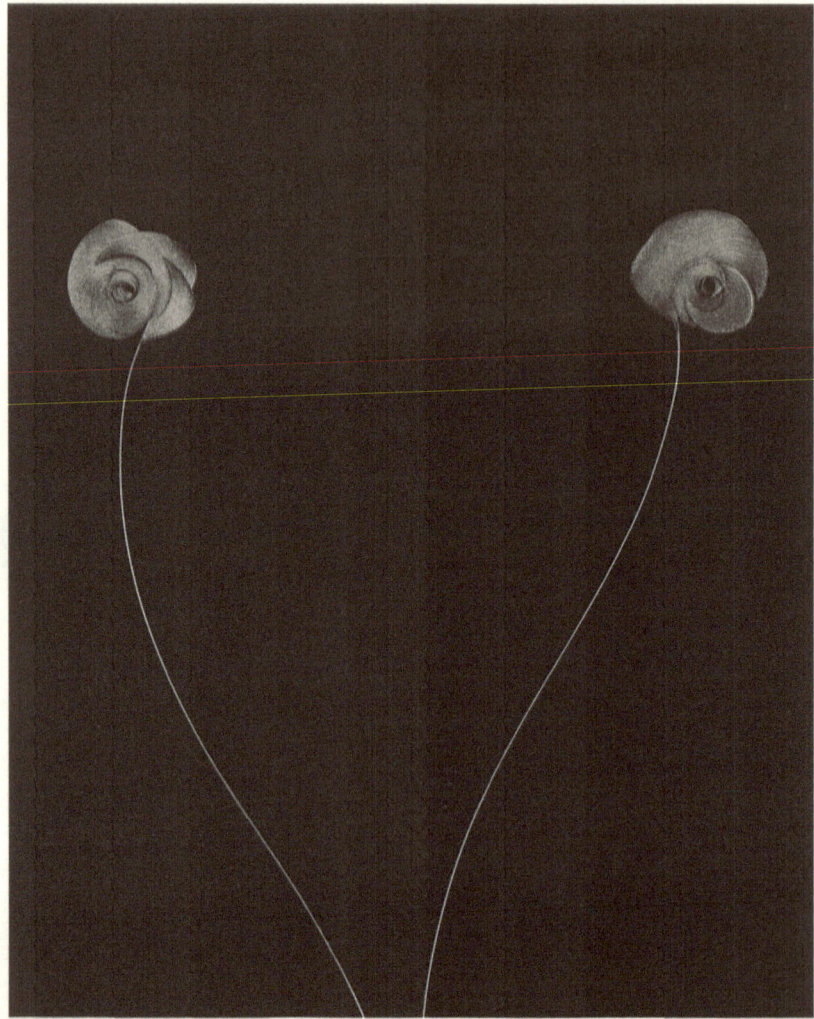

FIGURE 7. Outer tail feathers of a male king bird of paradise. © Kenji Aoki. Used with permission.

Dancing with the Birds showcases the twelve-wired tail of the bird of paradise, the nine-step dance ritual of the Carola's parotia, and the four-foot-high architectural masterpieces of the bower bird. Adaptationists argue that such grandeur signals the "vigor" of male birds, considering the calories males must expend to perform these dances and the time bower birds spend searching for the perfect ornaments for their bowers. The males with the most ornaments signal the most vigor and thus their evolutionary advantage over other competing males, so the adaptation argument goes. But Brennan witnesses these displays, these dances, and, she chokes out, laughing, "I don't see vigor there!" Moving her arm over her face to mimic a common mating move performed by male ducks (a display not present in the documentary), Brennan says, "Some of their moves are like, 'I'm going to stick my beak under my armpit now.' . . . I'm sorry, that's not vigorous. . . . It's sexy. 'Look at me!'"[44] On display is allure, beauty, *the bizarre*.

Brennan defines "the bizarre" as that which cannot be predicted or expected, that which fails to "make sense" or lies beyond the bounds of scientific knowledge and brings one to the brink of critical imagination.[45] Unbeknownst to me at the time of our interview, Brennan had developed this relationship to the bizarre after meeting Katy Payne during her years as a graduate student at Cornell. At that point, Cornell had been dubbed the "Temple of Adaptation" by Stephen Jay Gould. What Gould meant as an insult, biologists at Cornell would take with great pride, printing the title as a banner and displaying it throughout the department. Within the walls of the temple, though, lived observers like Payne, the exemplar of the scientist as witness, whose way of being in the world would influence dissenters like Brennan.

Brennan's memories have to do with the indelible impression Payne leaves on those who enter her presence, an impression that has to do with her way of punctive listening to the surroundings of her environment. Brennan described Payne's "way of being"—not having heard my own thoughts on the subject yet—as one of unwavering attention: nothing exists outside of what is directly before Payne. It is this element of Payne's praxis that most altered Brennan's work. "I remember those stories about how they figured out whales were singing, how she figured out elephants were communicating in this infrasonic way," Brennan remarks, "and those descriptions were about paying attention. . . . Just being in a moment where you're so *tuned in* that you find something that others have missed."[46]

Even more impressionable to Brennan was how Payne's comportment, or mode of attention, reflected in her way of relating to people. Finding out about my experience with Payne, Brennan asked, "Didn't [Payne] make you think . . . or make you *feel* when you were with her that you were the most important person?"[47] Payne is not someone who would check her phone or allow her focus to be distracted, Brennan reminisced. "She's just like *this*," Brennan whispered, putting both hands by the sides of her eyes and then moving them toward mine:

> She's connected. She's paying attention to you. She's listening to you. . . . I don't know if the person was informing the science or the science was informing the person . . . [but] that's certainly one of her most amazing qualities. She always makes you feel important and . . . that kind of intensity of connection . . . is part of what, I think, made her find the things she's found.[48]

Brennan did not say that having witnessed Payne's punctive listening was what allowed her to depart from the rigid traditions of the adaptationists. There is no denying, however, that Brennan is now far removed from the temple, chasing after the bizarre, and doing science with the same "social justice-y" bent as Prum.

In 2013 Fox News reported on Brennan's grant from the NSF to study the evolution of duck genitalia as evidence of "wasteful government spending." "It's part of President Obama's stimulus plan, and it's just one example of the kind of spending decisions that have added up to massive debt and deficits," Shannon Bream reported to viewers. Sean Hannity then quipped, "Don't we really need to know about duck genitalia, Tucker Carlson?" to which Carlson replied, "I know more than I want to know already!"[49] Fox News would go on to poll viewers that day about whether they believed Brennan's study was an example of wasteful government spending—88.7 percent would say yes.[50]

In response, Brennan would publish an article in *Slate*, arguing that conservative news sites miss the point of "basic science." Not all scientific research produces "payoffs" that are immediately apparent, and the goal of basic science is to perform incremental work that may later be synthesized into wider-ranging applications. Instead of arguing for the implications of her work—which would come later—Brennan asked that individuals learn to listen, both to each other and to the happenings that surround them in their natural environments:

> Generating new knowledge of what factors affect genital morphology in ducks, one of the few vertebrate species other than humans that form pair bonds and exhibit violent sexual coercion, may have significant applied uses in the future, but we must conduct the basic research first. *In the meantime*, while we engage in productive and respectful discussion of how we envision the future of our nation, why not *marvel* at how evolution has resulted in such counterintuitive morphology and *bizarre* animal behavior.[51]

Brennan's call for time—time to marvel, time to listen—speaks to her practice of punctive listening, of staying in a state where she can observe "the bizarre" until it begins to speak its own meaning. Demonstrated, too, in her research is the fact that wherever Brennan directs her attention, she seems to find what others have missed. Brennan humbly remarks: "I'm finding it because [nobody is] looking."[52] Now, I am convinced, no one is looking in quite the same way that she can.

The Punctum's Sensorium

Scientists studying avian worlds may not yet understand *how* birds understand color, dance, or ornamentation more broadly, but they do know *that* birds respond to these aesthetic elements in social ways. As the biologists throughout this book have demonstrated, studying aesthetics in natural environments requires a certain comportment of self that opens individuals to punctive encounters, or to the affective symbolicity that Langer defined as inaccessible via discursive modes. The indefinability that makes the affective symbol problematic for discursive theories is precisely what opens the scientist to encounters where punctive listening may take place. And whereas most of this book has focused on how humans receive and interpret those punctive encounters, acknowledging animals as symbol users is perhaps most necessary for biologists like Brennan and Prum, who delve into why animals make the choices they do. Indeed, Prum's aesthetic evolution centers almost entirely on acknowledging that animals we do not regularly consider symbol users may traffic in symbolicity more than that for which we give them credit. Put simply, aesthetics means something to animals.

The rift between aesthetic evolution and adaptationists is even reminiscent of debates within semiotics circles concerning differing definitions of the symbol and how human language differs from methods of communication outside of human cultural systems. Recall anthropologist Eduardo Kohn's proposition that animals possess the ability to "icon" and "index" but not "symbolize." Kohn's argument is that animals can understand representations of likeness and metonymic correlations, but they lack the ability to understand the conventions needed to process indirect representations built through sociality. Adaptationists seemingly align with this argument of Kohn's: claiming that ornamental displays signal a male's fitness is to suggest that animals understand and communicate via signs that draw direct correlations (beauty = fitness). Prum's aesthetic evolution, on the other hand, speaks to a semiotics that expands beyond correlatives and includes a capacity for desiring, pleasuring in, and responding to aesthetics, one usually considered as an exclusively human capacity for making and being moved by symbols.

To clarify, Prum never explicitly argues for such a semiotics; its remnants, though, rest between the lines of how he describes beauty in nature. Rather than an expertise with symbolization, Prum has a predilection for uncovering how aesthesis has accrued via evolutionary processes. Consider how Prum explains discovering the colorful plumage of dinosaurs: "Discovering the color of a dinosaur is more than just fun; it raises a host of fundamentally new questions [concerning how] the dinosaurs coevolved to be beautiful—beautiful to dinosaurs themselves—long before one exceptional lineage of dinosaurs evolved to become flying birds."[53] "Beautiful to themselves" reminds one of "art for art's sake" arguments, but need not circle aesthesis back to theories prior to affect and the social construction of knowledge. Instead, we need only place Prum in conversation with theories of culture in animals to realize that current understandings of semiotics do not allow us to understand animals' relationship with aesthetics. Placing Prum's questions of aesthetics alongside Kohn's posthuman semiotics shows another example of a scientist searching for ways to suspend symbolicity and "sink into" nonhuman animals' ways of knowing, this time to understand how it is that animals come to be "beautiful to themselves." Suggesting that nonhuman animals may make their own semiotic jump into the realm of symbols.

Langer would argue for an expansive symbol, and Prum's work furthers the idea that the more than discursive has meaning through how the senses do their own rhetorical work. If *logos* is the dominant mode of human rhetoric, *alogos* is that mode extending into rhetorics beyond the human, according to rhetoric scholar Debra Hawhee. *Alogos*, Hawhee explains, is sense-inflected. Theorized at least since the time of Aristotle, *alogos* named a part of the soul "that operates without reference to rationality; it is nonrational (as opposed to *ir*rational)."[54] Using Langer's multiple valences of the symbol to positively fill in the work of *alogos* presents an elucidated understanding of what happens semiotically in the sensorium of the punctum, when birds are moved by their subjective, aesthetic tastes, when humans are moved by the beauty of birds, or even when human scientists and their nonhuman subjects symbiotically engage with one another. Affective symbols are symbols more immediately and perhaps exclusively known through the body that may be shared by both human and nonhuman animals.[55]

However, considering only Prum's work on this issue is misleading—dancing manakins and extravagant birds of paradise are almost undeniably beautiful. The same case for beauty is not as easy for neuroethologist Ron Hoy. Hoy studies spiders—jumping spiders to be exact—of the *Habronattus* genus. Biologists have determined that the *Habronattus* spider is the "bird of paradise" of the spider world, what with their bodies of vivid blue and red in patterns that look more like quilts than arachnid bodies. Also known as "the peacock spider," *Habronattus* spiders challenge the common fear humans have of those with spindly legs who crisscross our bodies and deliver potentially poisonous bites. Arachnophobia even populates the list of top ten common fears in every culture worldwide that has been surveyed, Hoy reports. Showcasing the beauty of the *Habronattus* spider would overturn such fears, or so Hoy believes. He tells me—in all seriousness—that his work may elevate the spider to "human status," much in the same way that we consider whales and elephants.[56] We deem whale songs beautiful, full of rhythmic, musical pattern. We call elephant behavior emotional, filled with markers of joy and mourning. But we still just want to stomp on spiders.

FIGURE 8. *Habronattus* jumping spider. Used with permission from Jürgen Otto.

Maybe it would help, Hoy suggests, if we saw *Habronattus* spiders as dancers. Not dancers like the manakins, who craft elegant, orchestrated routines. *Habronattus* spiders have much more funk, antennae flung overhead, like human arms waving in the air to "Y.M.C.A." by the Village People.[57] Clearly, Hoy remarks, these dances are "performances," dance-offs, so to speak, that hint at aesthetic evolution in spiders as well as birds. In insect as well as avian worlds, it seems, beauty happens, and what comes after in evolutionary history is a by-product of individuals' subjective aesthetics.

Yet Hoy has never said as much in his scientific writing. These discussions only happen in the lab when he and his colleagues place bets on which male the female *Habronattus* spider will choose. "If I'm watching an experiment," Hoy clarifies, "I [can] say, 'I bet she chooses him.' If I were a female jumping spider, [I would] because look at the extravagance of the moves! Look at how high he stands on stilts, on his legs. . . . That's beautiful. That's a really great performance."[58] Writing up this research, however, beauty must not be named—objectivity must enter in that late stage.[59] There is a list of checkboxes, too, scientists must consider when they sit down to write about aesthetics in evolution, Hoy reminds me, and "if you're too quick with the attribution of beauty . . . you're going to get swatted down." So Hoy avoids the subject, and he understands where Prum's critics stand: beauty is a higher-order emotion that "requires having a mind of comparison."[60] To call something beautiful is to define something by other things you have encountered and remembered. Acting on feelings of beauty requires not only decision-making abilities, but also meta-cognitive abilities, or the ability to act on what is remembered. Hoy has spent his career proving that such abilities exist beyond the human, in crickets, mosquitoes, and spiders. The *Habronattus* spider, for example, has route memory, or the ability to maneuver around barriers via spatial mapping.[61] Such "memory" may be stored in the body and not necessarily contingent on rational thought, but saying that animals interpret beauty is saying that they have the ability to recognize and act on nuance, as beauty is unable to be defined, or can be defined only through a stored catalog of subjective experiences.

Hoy is careful to point out that scientists cannot prove animals' abilities to "label nuance and be able to reach for it." As far is known, that ability is human and linked to verbal capacity, Hoy explains.[62] Here is where Prum would disagree. Beauty need not be linked to neural correlates in order to be proven; instead, scientific methods need only value experiential knowledge. And *that's* what his critics refuse to say: Prum falls under criticism because taking his theory of aesthetic evolution seriously requires ushering in new knowledge and value systems into the sciences. Oddly enough, in *Evolution*, Prum articulates knowing in multiple ways via multiple registers through language itself. Noting that the study of animals' subjective experiences requires a different epistemology, a way of knowing that has less to do with detached understanding and more to do with sensorial and affective immediacy, Prum remarks upon the limits of the English language. Because there is only one English verb for "knowing," scientists are unable to imagine the subjective experiences of their subjects of study or even come to terms with their own subjective experiences encountering those subjects.[63] Other languages account for a wider range of experiential knowing: "In Spanish," Prum explains, "to know or understand a fact is *saber*, but to be familiar with someone or something through experience is *conocer*; in French, these verbs are *savoir* and *connaître*, and in German *wissen* and *kennen*." Birding and, by extension, good ornithology "is about knowing in [this] more intimate, profound way," through this experiential kind of knowing that occurs by listening to nature's punctum.[64] Encountering and acknowledging this punctive, affective symbolicity may be one of, if not the initial, rhetorical moment in good scientific work.

Because the senses are always examining the surrounding world as the basis upon which individuals form discursive interpretations, the brain codes sensory experiences according to one's rhetorical framework, as I have elsewhere argued.[65] To engage in punctive listening, then, scientists must go

SWALLOWED BY BEAUTY   127

beyond the logical limits of their rhetorical framework. Opening to a broader rhetorical system allows us to value ancestors beyond the human, a tradition much older than Prum, Brennan, and Payne.

Being Swallowed by Beauty

Reaching the limits of discourse, and then really beginning to listen to nature's impositions—what we have come to call punctive listening—is ushered in for Prum and Brennan through feminist, posthumanist praxis. To put it another way, punctive listening involves decentering human knowledge as the only or best way to know and encounter the world, which is exactly what Prum and Brennan had to do when they were confounded by the morphology of duck genitalia. The evolutionary history behind that anatomy was at first "inaccessible to [them] because the only way to analyze it was to explain it away rather than actually recognize the thing to be explained." The disconnect, Prum explains, was that he was trying to explain semiotics in natural environments exclusively through the framework of disciplinary, discursive symbols. The sounds and semiotics in natural environments, Prum soon realized, could explain themselves: "It [took] recognizing my own subjectivity to recognize this.... [T]hat *posthuman move* meant that ... I don't have any special authority to explain the birds. Let's let them explain themselves."[66]

When I ask Prum how he came to trust personal, subjective experiences as integral to reputable scientific study, he talks about growing up as an avid bird watcher. From an early age, Prum learned to trust his senses. Just by looking and listening, he explored a world different from our usual distracted forms of processing; he engaged with a different semiosis in the field. By middle school he had become such a skilled birder that he had to seek out experts to confirm his sightings. One by one, Prum's experiences were confirmed, and it was the "success of those [subjective experiential] acts one at a time ... [the] first-person [recognition] that means *I* have this experience and knowledge that I can acquire from the world" that brought him to trust punctive knowledge.[67] The reputability of "first-person" experience came to outweigh anything that Prum could learn in a textbook. Still, when findings fail to match up with established scientific fact, Prum trusts his senses, that knowledge that comes to the body through punctive listening.

Once when Prum presented the idea of being led by beauty and thus understanding a different valence of beauty in evolutionary biology through this new material and posthuman epistemology to a colleague, they responded in despair: "That's nihilism!" Such a reaction is quite common among colleagues, Prum reports. If his thesis is correct, then many presume that "the universe is not rational." And if the universe is not rational, or not rational in a way that we can make "rhyme or reason" of, then scientists adopt a "bleak worldview" that deprives them of "any sense of purpose or meaning in life."[68] In Prum's experience, once humans lose the ability to explain the universe, they feel they have also lost their place in it. This displacement and sense of insignificance is exactly the point for Prum and Brennan. That is what makes beauty in our natural environments—in all its arbitrariness—worth attending to, revealing how we human animals almost were not. Calling us to our fragility, Prum argues, "Our mind is not . . . *anything special*. . . . It feels special to us because we have [it]. But everything about it is . . . a humility that that's just an accident."[69] The magnitude of humans' happenstance becoming, Prum hints, is where sound scientific theory should begin.

Much of scientific inquiry, to be fair, begins with the senses, from feelings of awe and aesthetic wonder. Prum simply calls scientists to "be honest about this aspect of our inspiration."[70] Should scientists do so, they may more easily access the humility it takes to suspend disciplinary interpretation and symbolize alongside their subjects of study. Payne described such a comportment of humility as "clinging to a cork" in the middle of the ocean, a position that makes it difficult for her to feel as if she—a human animal—has any real authority over the natural environment. The need for a bodily comportment that leads to punctive listening, the quieting of human symbolicity, resurfaces when Prum posits "letting birds speak for themselves."

It is Brennan, however, who best describes not only this comportment in personal encounters with her natural environment, but who narrates how that comportment is inspired. It comes when humans confront the magnitude of their natural environment and its force, a feeling Brennan describes as being "swallowed" by the vastness of what is before you. To study life in its becoming and on its own terms, it seems, the scientist must be swallowed. Brennan was first confronted by the magnitude of nature while conducting research in the Amazon. This same loss of self into the vastness of space persists, without fail, in the present day whenever she returns to the rain forest. "You step in on that forest," Brennan describes, "and you are immediately faced with the reality of just how big the world is and how amazing nature is and how insignificant you are. *The force could swallow you*, and nobody would even know you were there."[71] Brennan never conveys fear of being swallowed. This is no confrontation with the void at the edge of an abyss. This is more like slowly falling back into the arms of strangers who catch you in a "trust fall." Falling, you become a different person from the sure-footed individual you were just moments before. And falling into the arms of others, you realize that those who catch you were never strangers in the first place.

Yet Brennan's expression of being "swallowed" is not only the feeling of falling, but of falling into insignificance. This particular posthumanism is one in which the human dissolves into a larger whole and theorizes from that place.[72] Not quite a marginal or decentered perspective, being swallowed is closer to a loss of perspective that turns an individual to listening when the magnitude of what is before them exceeds their ability to rationally interpret all of it. Being swallowed by the force of nature disrupts logic and symbolization and brings one into the realm of punctive knowledge. Left in the realm of affective symbols, scientists—as witnesses—value knowledge that floors us, that pushes through disciplinary traditions and stops us in our tracks.

For Brennan, being swallowed transformed her work, teaching her to punctively listen, to find the "weird" or the "bizarre" that emerges in nature. The one reliable element of her research is knowing that she will "find something weird wherever I look, because it's come through every time." Brennan proves time and again that nature is weirder than that for which scientific empiricism gives it credit. She began to suspect as much after she dissected her first male duck and realized that duck penises made no morphological sense. After watching tinamous mate, she knew that studying duck penises would involve quite the learning curve, but she could not have guessed how lymphatic they were, how oddly shaped, how long. She recalls: "The first thing that came to my mind [was] 'Oh my god—this is *so bizarre*. This organ is so bizarre; it's unlike [anything] I had ever imagined.'" After this, Brennan quickly moved to dissecting female ducks; seeing the male duck penis convinced her that the females must possess "something weird" anatomically as well. When she found female ducks' "labyrinth-looking vaginas," Brennan remembers, "it was like mind blown."[73]

Brennan would spend nearly a decade studying this bizarre duck genitalia, but another day would swallow her anew: the day she dissected a dolphin vagina. Recounting that moment, she motions to a bright blue vaginal mold on her bookshelf: "That's not a duck vagina. Even though it looks pretty much like a duck vagina. It's got that big pouch and those spirals. [But] it's a *dolphin* vagina. . . . When I pulled that thing out of [the dolphin], I almost died. . . . Because the ducks were weird! And now the dolphins are weird, too?"[74] That's how Brennan concluded that these seemingly rare genitalia were "not really weird. It's just that nobody had looked." Ducks and dolphins are common animals that have been extensively studied for decades. Her findings, then, are "commonplace."[75] They only *seem* weird because researchers did not look at female morphology before with enough regard, or perhaps could not make such anatomy mean anything according to hegemonic narratives of their disciplinary field. Either way, Brennan began to articulate structures in natural environments previously unknown to the sciences, and she did so through punctive listening, through being swallowed by how nature speaks on its own terms.

In her teaching, Brennan spends time considering how to bring students into encounters with the weird, the bizarre, and the beautiful, or how to make students feel the magnitude of nature in a way that takes them to their limits, intellectually and sensorially. One way she sets the stage for such encounters is with a "Termite Lab." Students are given different colors and brands of pens and asked to draw different lines across a large white sheet of paper. After students have doodled, Brennan drops termites onto their color-filled sheets. Termites scurry across the pages, as they are apt to do, but before long, they begin to follow only certain lines on the page. Students watch as termites fall in line, charting a very particular course. "How *bizarre!*" her students yell. Brennan then leaves them with the responsibility of determining what is going on, how the termites have charted a course, how they are able to follow one another, why they follow some lines instead of others. Maybe it has to do with the thickness of lines, students posit. "Test it," Brennan replies. Maybe they follow straight lines instead of curly ones. Test it. Maybe they prefer certain colors over others. Test it. Eventually students argue that one particular brand of ink attracts termites, after which Brennan reveals that Bic pens contain a certain chemical that mimics the pheromones termites will follow. Brennan could have just told her students about this fascinating occurrence or shown them videos of the experiment, but then, she worries, "it would be just another forgettable factoid, right?"[76] Giving students this experience allows them to be swallowed by nature's punctum.

When I ask Brennan if she had a similar entry into the field, she shakes her head and—oddly enough—tells me a story about a whale. Brennan had been working as a research assistant on a marine vessel tracking whales and dolphins around the Galápagos, funded, as fate would have it, by none other than Roger Payne. At that time, she had no intention of getting a PhD and becoming a biologist. She was simply tracking wildlife because she needed a job. One type of whale, the beaked whale, proved particularly difficult to track because of their behavior patterns: diving deeply, surfacing for only a quick breath, and then returning to the deep. One day, however, Brennan spotted a female beaked whale with a baby. Because the female had her baby, she was not diving. Unable to identify which whale this was, Brennan took photographs and sent them to an expert at the National Oceanic and Atmospheric Administration. She would soon learn that this sighting was likely the first ever of a Longman's beaked whale, an animal who had before only been described on the basis of its skull. She was swallowed: Here was a creature nine feet in length weighing over 2,500 pounds and scientists had *missed it*? "How many other things are there," she asked, "that we just don't know?" Prior to that day, Brennan had no sense that there was "a whole world to discover of things that we don't know.... [I thought] there were just ... very specific things that we didn't know.... I thought that science was a lot harder than that."[77] Since then, Brennan has remained preoccupied by how much "we just don't know," by nature's magnitude, by how its force punctures through our discursive ways of knowing. Payne once mentioned that when Roger Payne read *Moby-Dick*, "he just swallowed [the book]," but Brennan brings me to wonder: Was it Roger Payne who swallowed *Moby-Dick*, or was it the whale who swallowed him?

Speculative Science

Arguably, we could look at Brennan, Prum, Payne, and their methods and say that they are radical empiricists, scientists seeking knowledge through sensory experience, the practice reflected throughout histories of science and that purportedly still guides scientific methods. The reception of their work, however, is telling: either this is not "empiricism" and thus not deemed reputable, or this *is* empiricism and contemporary scientists are allergic to it. Jerry Coyne, for instance, argues there is no proof in Prum's speculative science.[78] The problem, it seems, is not that their empiricism begins with the senses, but that it stays in that affective space, in the open semiosis where animals' individual subjectivities guide the work. And whereas Prum and Payne leverage their privilege to fight for the importance of their findings, Brennan reminds us that such acceptance is conditional. Even Payne herself was never invited properly into the academy, all the while young scholars built their careers on the foundation of her findings. "[Payne's] methodology is not compatible with what we ask people [to do] in the academy," Brennan remarks. "That's not to say it's any less important or valuable, because I don't think it is. But the reality is that academia would not have made a space for [Payne]. It wouldn't."[79] Brennan reports that Payne's description of herself as an observer rather than a scientist is more accurate.

Oddly enough, part of this practice of observing as a witness, of punctive listening, involves a more radical practice than Payne herself would allow. That is, Payne had corrected my anthropomorphic tendencies; however, these tendencies were not only shared by Brennan but embraced by her. Asking the question "What does *she* want?" requires asking how the female under study feels, Brennan maintains. "I *do* anthropomorphize in the sense that I put myself in the feet of the female duck," Brennan admits.[80] Brennan goes even further to say that such anthropomorphism cannot so easily be removed from her work once she moves to publishing her research. Because identifying with the individuals is central to her thinking process, the scientific conclusions at which she arrives are imbued with such anthropomorphic tendencies.[81] This practice may keep her from being "as objective as [one] might be," but she contends that being more subjective is not necessarily problematic. Anthropomorphism, Brennan explains, serves as a useful heuristic that opens scientists to more expansive lines of inquiry, as long as they stay committed to listening to beauty.

Anthropomorphism becomes problematic when it limits explanations, when scientists presume to know what is going on in the mind of an animal because of what is going on in their own minds. "We can't know what's in the mind of an animal," Brennan explains, "so if we put something there that's something the way a human might interpret it, we might miss alternative explanations that we haven't even considered because we have a limited human experience."[82] But using human experience as a base subjectivity through which to understand animals' subjectivity may be productive, leading scientists into nonhuman ways of feeling and knowing the world. This idea of human animals thinking with and alongside nonhuman animals echoes Hawhee's articulation of the two as "partners in feeling," relationships "carrying lessons, calling forth, engaging, and shaping humans' beastly sides."[83] Perhaps, Brennan conjectures, female ducks lack nerve endings in their vaginas that make the male duck's penis not as painful as it might seem to us. Brennan's anthropomorphism thus brings Payne's critique into relief: feeling with the animals is apropos of felt objectivity, but using human experience to interpret what is not present makes the problematic slip into hungry, extractive listening.

For Prum, theorizing through anthropomorphism and studying nonhuman animals' subjective experiences brings him frequently to speculate. The accusation of speculation, which would shame most scientists, is one Prum embraces because it is central to a felt objectivity that resists causal closure and allows individuals to make meaning of their own. Think of speculation, Prum prompts, as "idea generation":

> What I didn't want people to think [while reading *Evolution*] was the idea that I think this is *done*. Rather that this is opening a new area in science. That's what I really wanted people to realize: when I say "speculative," that's what it is—I'm not done. This is just the *creation* of [what might be considered] a queer evolutionary biology. Just a *creation* of the idea of autonomy as a force in the origins of humans. Or just *opening up* the prospect of pleasure as its *own* cause of being.[84]

In a way, Prum is enacting his theory of aesthetic evolution. If what becomes of lineages biologically, or here intellectually, is based on the arbitrary rather than the determined, then speculation is what opens up a new iteration of that lineage, new possibilities, new becoming. Further, if these speculative and anthropomorphic methods signal the erosion of objectivity as we know it, they need not signal the end of "science." Instead, it is the start of good data. When objectivity goes, all you are left with is "quality," Prum argues, quality determined by "opening up your data" and allowing other people to test and evaluate it.[85] Speculation, then, is a valence of scientific inquiry that requires a double openness: an open invention process of idea generation and an open presentation of one's work so that others can evaluate what is being used to sustain it.

Lost on no one, though, is Prum publishing about speculative methods through feminist and posthumanist epistemologies in a way that Brennan is not. Prum, at least, is open about these gendered discrepancies in STEM fields, about how women must protect their careers to earn tenure and promotion in a system that is still extensively gendered and racialized. This reality—that certain bodies are called to be more objective than others—is one that Brennan, as a woman of color, has lived with throughout her scientific career. It is part of the reason that she chose to change her last name when she was married. Formerly Patricia Rodriguez, "Brennan" offers her field a name that leads many to assume that she is of Irish descent. Recently a study confirmed Brennan's hunch: when researchers distributed eight identical curriculum vitae to biology and physics professors at various large public US universities but changed the names on the CVs to manipulate race (Asian, Black, Latinx, and white) and gender, Black women and Latinx women were rated lowest in being hireable.[86] When Brennan came across the study, her reaction was one of victory. "Hell yeah!" she remembers yelling, because Brennan has known how others mistake her positionality for quite some time. Brennan tells me how she has been invited to speaking engagements by people who only knew her by name. When they introduce Dr. Patty Brennan "and I walk in . . . they do a double take."[87] She laughs knowing those people wonder whether there has been some mistake. No mistake, "because I'm Patty Brennan, Irish girl"— she pauses and smiles—"until I'm not."[88]

Both Brennan and Prum situate themselves, then, to practice this felt objectivity in the academy, Brennan through the power of perceived whiteness and Prum through his "whiteness, Ivy-ness, [and] gray-beardedness."[89] The stakes of their argument for female autonomy are too important not to fight for this method. Their work stands to change what science is, who can engage in scientific inquiry, and what makes a published scientific study defensible.

Good science is no longer the most modest science, as they show. It is science that holds itself accountable to others, to the positionality of human animals and the subjectivity of nonhuman ones. When rhetoric is brought to bear on the sciences, we often point out its flaws, but part of our imperative is also to uncover how we might help scientists move toward this humanistic accountability.

Encouraging along these lines is Prum's conviction that this scientific praxis is one that everyone may have access to at least in some degree because it is "not synonymous with hypothesis testing." Prum explains this broader science:

> Galileo, five hundred years ago, took a telescope and looked up at the heavens, and over a series of nights, he saw that . . . the moons of Jupiter were orbiting around [the planet]. And with a *simple logic* proved to himself and then to the world that we were not the center of creation. And he was not testing any *damn hypothesis*. He was engaging, [looking] at the world. And . . . taking my binoculars, my optical devices, and staring into the lives of birds, I aspire to make the same kind of discoveries in my work. . . . [Like this, we engage] with nature, and mak[e] discoveries about the things for which no controlled experiment is even plausible.[90]

Everyone who wants to ask these "simple logic" scientific questions can. "Each person on the planet sees a problem and will ask a question based on their individual experience," Brennan says. "That's why there's got to be room in science for someone like me . . . because if there's no room, then these questions don't get asked."[91] It is the argument that warns about what happens when certain white, racist norms become dominant: the sciences must include wider ranges of sensing, feeling, and knowing if these disciplines are to encounter all the natural worlds of which humans are a part. The practice of objectivity, too, cannot be singular. Its multitudes manifest differently at different moments in scientific practice, and its subjective valences, rather than being ignored or dismissed, Brennan and Prum show, can be resourceful, if not integral, to scientific knowledge. What's more, the truest objectivity—arriving at a place where one can feel and listen to the punctive impositions before them—may come from being swallowed.

Beauty Happens

By now we know that scientists encounter beauty when they listen to whale songs or watch manakins dance, and we've heard that they are taught to leave those experiences out of their scientific articles. That maxim—you cannot talk about beauty in science—is one Hoy has heard his whole life. But beauty is what brought him to biology, or, at least, beauty will be his legacy. If Hoy has it his way, he will "live long enough to build a bridge between embodied markers of reaction and what happens to a brain" when it experiences beauty.[92] Hoy, who began his career in the 1960s, will not guess at the standard of beauty for the jumping spider or the rat, but intuitively he knows that such a standard exists. John Durham Peters extends the same idea when he remarks that "all life feels beauty, however inchoate." He continues, "Our sense of beauty is not just some kind of cultural imposition, but a deep part of the infrastructure of natural history."[93] And as Langer, Payne, and others have taught us, being able to map the neural bridge between aesthesis and the neurotransmitters that measure beauty is but one form of knowing, and not the form with the most feeling.

Nevertheless, Hoy's wish for the close of his career gets at the heart of Prum's inquiry and the question of this entire book: Can beauty change science? Arguably, it should. Hoy is not alone in this wish; it is also the reason botanist Robin Wall Kimmerer entered her scientific studies. Yet when she told her adviser this idea directly—that she "wanted to learn about why asters and goldenrod looked so beautiful together"—he retorted, "I must tell you that *that* is not science. That is not at all the sort of thing with which botanists concern themselves."[94] He enrolled her in a botany class to prove his point. That adviser, Kimmerer reports, made her doubt where she came from and what she knew in the implication that his way was the only *right* way to think. Kimmerer, however, never let go of the idea that her question was one that science would benefit from. As she puts it:

> Together, the visual effect is stunning. Purple and gold, the heraldic colors of the king and queen of the meadow, a regal procession in complementary colors. I just wanted to know why. Why do they stand beside each other when they could grow alone? Why this particular pair? . . . What is the source of this pattern? Why is this world so beautiful? It could so easily be otherwise: flowers could be ugly to us and still fulfill their own purpose. But they're not. It seemed like a good question to me.[95]

Kimmerer spent years studying botany, becoming a scientist, and never losing the idea that science *could* be about beauty, that it *could* be about "the embrace between plants and humans." Why does no one ask nature, she wonders, "What can you tell us?" How does beauty work? What can you teach us? Instead, she argues, science reduces individuals in nature—plants even more than animals—to objects, not considering them *subjects* to be listened to and learned from.[96]

If we take Kimmerer's example and watch asters and goldenrods long enough, the bees teach you why they are so beautiful, why they grow together. See, the bees dance between both, drinking nectar and gathering pollen from each. "It is this dance of cross-pollination that can produce a new species of knowledge, a new way of being in the world," she says. She further suggests that the "September pairing of purple and gold is lived reciprocity," which means that they teach us how to be together. "The wisdom [of the asters and the goldenrods] is that the beauty of the one is illuminated by the radiance of the other." She thus asks, "Science and art, Indigenous knowledge and Western science—can they be goldenrod and asters for each other? When I am in their presence, their beauty asks me to respond, to be the complementary color, to make something beautiful in response."[97]

Losing touch with beauty in science has changed our relationship with nature, has changed what we produce and study and who we hold ourselves accountable to, which is to say: Losing touch with beauty in science has changed our relationship to knowledge itself. We have made knowledge too narrow. I have spent the last four years interviewing researchers like Prum, Brennan, and Payne, and what I've learned most is about listening. About entering the field *without* preformed questions or hypotheses. About learning the names of whom you'll meet on walks in the woods. About walking the same trail as often as you can to see how every individual on that trail changes with each passing day, just as you do. Blame it on naturalistic fallacies or sensory bias, but that is the real reason science is still allergic to beauty: waiting for it, listening to it, being quiet with it, takes too much time. And in the end, there is no guarantee of what we might find.

Conclusion

Truth be told, it was Ron Hoy, the spider scientist, who sent me down to Oak Grove, Louisiana. After our interview, he stopped me as I rose to leave his office and asked, "So, what's *your* story? Why do *you* study beauty?" When I went to stop the recording before answering, he asked me to keep the tape rolling—saying that I may need to listen to my answer later. Maybe because Hoy had spent time in coastal Louisiana, studying how crawfish claws regenerate and how mosquitoes listen, I shared with him my history with hurricanes. How I missed months of school the first time, how, when we returned to share the place with our rivals to the south, I was interviewed by a writer from New Orleans who saw a story in me I never would have otherwise noticed. How when the winds, storm surge, and rebuilding happened again three years later, that same reporter returned, asking me to explain what was happening to us. How, even though I was a first-generation college student who had few ambitions in the arena of education, I spent the next ten years studying the ineffable and the limits of the sensible to understand why I am still unable to answer his questions. How I struggle to see a new landscape without imagining its trees with bare limbs, its homes with blue tarps, its streets waterlogged. How I find myself studying beauty in natural environments to reintroduce myself to my hometown, to finally accept that hurricanes, more than the people in it, were always what would define that place.

Never had I been so honest with the biologists in this study. In kind, Hoy invited me to linger longer and continue our conversation. He, too, was a first-generation college student, and now that we were being more honest, he needed to correct an answer given earlier in the interview, the only question he had deflected. I had asked about Hoy's earliest memories experiencing beauty—did his answer have to be about nature? If not, he would offer a different experience: Loretta Lynn kept coming to mind. Not only Loretta Lynn, but Patsy Cline and Ernest Tubb. Ask him about beauty, Hoy remarks, laughing, and all he can think about is "hillbilly music, country music" that he grew up hearing in Walla Walla, Washington. There, in its own brand of rural America, the instrumental music he would grow to love in college were absent on the airwaves—everyone preferred Loretta Lynn. When he listens to classical music now, Hoy finds himself often moved to tears. There is something about feeling the "embodied markers all hit[ting]": the hair rising on the back of his neck, chills running down his spine.[1] All the physiological markers of what it means to experience beauty that he studies in insects, he feels. And those country music stars were formative in priming him to live in a "world of sound." In many ways, the effect of country music on Hoy speaks to his relationship to science, to beauty, to academia writ large because, like that reporter did for me, country music taught Hoy "how to be 'meta' very early."[2]

To be "meta" is to be self-inferential; for Hoy, being "meta" comes not from adopting "an observer's stance," but querying what it means to observe, and observing how you are being observed. From an immigrant family, Hoy did not look like his white friends growing up, and whereas country music served as shared common ground with his classmates, he was hyperaware that these tunes were part of his friends' culture, the one to which he was assimilating, never the other way around. So, Hoy would witness his friends attend to how he listened to country music. He witnessed his friends notice his parents running a Chinese restaurant, the only "niche available to them" then. "So, I grew up as an observer" of observers, Hoy remarked. "I was an outsider looking in to a lot of people who didn't know how to deal with people like me."[3] If observation is regarded as the initial step of scientific method, Hoy's account reinforces how observation is multidirectional, multiperspectival, multisensory, an aesthetic act that conveys more than discursive knowledge. Beauty emerges from this aesthetic scene not because of Hoy's musical taste, but in the memory of how that music taught Hoy to observe, witness, listen.

This approach to listening to beauty and understanding beauty as part and parcel to scientific method—if not epistemic methods more generally—is what we have called punctive listening, emergent listening, but may also carry valences of diasporic listening. As defined by rhetoric scholar V. Jo Hsu, diasporic listening is "a critical orientation attuned to reciprocities ignored or obscured by normative frames," a method that centers connections and community among "the many ways our lives come into contact and tension with others." Further, when connection and making kin is prioritized over defining meaning or articulating meaning based on what we already

CONCLUSION

know, listening opens up the "possibility of unknowing."[4] In the realm of the unknown, we must trust the more than discursive, or that which is conveyed through affect, feeling, emotion, intuition.

More than anything, the desire to find beauty in dissonance, to live with what you will never fully make sense of, is what each biologist I interviewed had in common. Listening to country music, the rain forest, or whale songs; feeling elephants rumble or watching manakins dance—beauty emerges, differently for Hoy than Payne, for Prum than Brennan, for Fournet than May-Collado, for Wrege than Hedwig, for me than you. Beauty always has emerged differently for each of us—never pushing us toward resolution but asking us to accept ourselves and our worlds in all their messy, dappled evolution. Maybe that is why I have done little to define beauty in this book, concerning myself instead with how we conceive of what we are unable to articulate, how we make meaning through "wordless" moments, how nature's punctum reaches out, whether in photographs, field encounters, or how we face one another. In galleries, museums, and labs, to be sure, but also in how we are part of that nature we seek to understand, beauty is a method that asks us to pay attention, teaches us to how to listen, prioritizes respect for and connection with kin.

Rhetorically, beauty convinces us to set aside what we think we know—to witness beauty in natural environments and trans-species encounters is to respect how other ways of knowing and communicating are just as valuable as our own. Deep, punctive listening to nature's beauty thus serves as an antidote to hungry, extractive listening in dominant science and invites future studies of how scientific programs and praxis may make room for the aesthetic, creative, affective work of prescientific invention. Such programs and practices must be multitudinous, considering that just as beauty is not singular, neither is scientific method. The different accounts in this book attest to this: different subjects warrant different modes of punctive listening. In Payne's descriptive science, in which the work was to record and describe the songs of the humpback whale and the infrasonic rumbles of the elephant, punctive listening was slow and incremental, taking months and years of patient, humble practice. In Wrege's and Hedwig's conservation science, punctive listening involved working with communities to learn about the complex context of African rain forests and relied on the elephants' communication strategies to provide improved methods for resisting extractive industries. In May-Collado's phylogenetic approach, punctive listening included considering how songs and sounds come down from ancestors perhaps never otherwise considered or known. In Fournet's, Brennan's, and Prum's feminist approaches to science, punctive listening insisted on giving voice to female subjects, on adopting a comportment that allowed us to be "swallowed" by forces and knowledges beyond ourselves.

Here, at the end, it may be helpful to articulate the interplay between listening and witnessing in scientific inquiry. There can be listening that occurs without witnessing—hungry, extractive listening that listens for certain answers or certain kinds of answers, listening that does not bear responsibility to or stand in relation with all we are amidst. By contrast, there is no witnessing that occurs without listening, I would argue. Witnessing that is not deeply embodied, grounded, and responsive to the natural environment in which it occurs, or witnessing that is primarily cerebral and visual, is the fictional, modest "view from nowhere," not a form of bearing witness. The scientist bearing witness acknowledges the aesthetic excess in natural environments—all that can be sensed but perhaps not said, even all that can be felt but not quite sensed—as epistemic resources. As the biologists in this study have shown, scientific inquiry can be stronger and more revelatory when researchers lead with listening, shy not away from the resources of aesthetics, and understand their role as witnesses who aim to connect with their subjects of study and allow those individuals to speak, sound, sing for themselves.

Recall the different valences between techniques and technologies, with technologies marked as durable and replicable, and techniques as more malleable and fleeting—this book has done little to challenge scientific method as a technology.[5] Scientific method as a technology should be equitably available to anyone and independent of the researcher who employs its praxis. Those techniques, by contrast, employed during scientific method do warrant adjustment and adaptation if those techniques are not responsive to their subjects of study, if they do not allow the punctive imposition of nature's beauty to affect not only the scientist, but science itself. As this book has contended, visual observation—arguably the most dominant technique in scientific method—may not always be the best approach. Subjects often reveal themselves differently when researchers are listening. And rather than "just" listening, punctive listening is about honoring beauty, respecting intuition, and then lingering with kin through sustained attention so that new ways of sensing and knowing can emerge.

In applied or clinical sciences, punctive listening would look different still, no doubt, perhaps employing community-led invention strategies prior to hypothesis formation or even allowing hypotheses to be instated directly by communities themselves. After all, are applied and clinical sciences not made for the communities in which scientists and institutions are situated? Regardless, future studies are warranted to first observe the broad range of listening practices across other sciences—whether applied, clinical, physical, mathematical, or ecological—and give special attention to *when* researchers turn to listening.⁶ Here, again, beauty may serve as a wedge for considering how scientists face the limits of the knowable and how recursive returns to those limits might challenge methods of prescientific invention to result in better, more equitable practices.

Throughout this book, I have argued that in the early stages of invention, starting with punctive listening, taking beauty seriously as a method, and attending to what emerges intuitively at the beginning of inquiry may broaden scientific endeavors, challenging whose expertise is valued, what methods are most objective, and how meaning makes itself known. Here, I should pause to acknowledge that this study has not been altogether distinct from Kuhn's original study of scientific paradigms, but this study of punctive listening has offered a new location for Kuhn's paradigms, not only in discursive communities, but in phenomena itself and how the body of the scientist receives that paradigmatic affect. That is, the idea of "paradigms" has rhetorical history— with the Greek *paradeigma* meaning "example" or "pattern," and the idea being that an argument succeeds when one thing can be compared analogically to a similar example. Kuhn's paradigms proceed when explanation of present phenomena can be supported by past scientific studies that discursively explain what we presently witness. But Payne's punctive listening teaches us that the paradigm may also be the humpback whale, resonating through bodily vibrations, showing a way of communicating that can then be applied to elephants. Paradigms thus require not just a knowledge of discursive disciplinary traditions, but also a rhetorical way of being in the field, a way of knowing that is responsive to nature's punctum.

In this sense, beauty asks us to attend to those remnants of paradigms that are articulated through feeling. Beauty pushes science to its edges, asking scientists to attend otherwise, to listen to the more than discursive, to value the vivid, affective symbolism that abides in natural environments. Perhaps the most important lesson that beauty teaches is that these paradigms exist despite us. At least for Fournet, that is what the calls of the humpback whale implore, that their calling has nothing to do with our human need to understand or make sense of its meaning. "You know what's happening right now?" Fournet asks. "Whales." In the depths of an ocean warmer than it should be, whales are, at least for now, "*still there*, still calling, still doing exactly what they do."⁷ As we inevitably impose, and in how we impose, may we remember that whales—and all other kin—warrant first our listening.

Acknowledgments

There are many worlds in which I never come around to writing this book. One of my greatest joys has been watching this work emerge from the stories, conversations, questions, and impositions from those I admire most. When I set out on this project, I aimed to consider how biologists study the ineffable in the field. I never expected whales, elephants, ducks, dolphins, jumping spiders—much less beauty—to fill the pages. To say that I am grateful to have spent the last few years learning from so much beauty, coming to terms with how we humans have the most to learn on the evolutionary stage, is an understatement. The biologists I met and the subjects they study changed the book and my career, but also my sense of what it means to tread on this Earth with so many others.

As I studied ways of relating to nonhuman kin, I did so as a settler scholar. Most of this book was written on the land of the Atakapa, Chickasaw, Coahuiltecan, Comanche, Cherokee, Ishak, Osage, Tonkawa, Seneca-Iroquois, Shawnee, and Susquehannock. Its research also took place on the land of the Cayuga, Golden Hill Paugussett, Quinnipiac, and Western Abenaki. Because it is not enough to acknowledge as much, I aimed to think with Indigenous theories and perspectives to help decenter my disciplinary training that primarily featured Western theorists and thinkers. Alongside feminist science studies and my home discipline of rhetoric, this reading challenged my notions of what science is, where science happens, who can be called a scientist, and what data can be considered credible in scientific inquiry.

Every page of this book is indebted to the biologists, composers, and audiologists who spent time sharing their stories and ideas with me over the years. Their time is valuable, and yet no one thought twice about sitting for hours and sharing vulnerability with a humanist. Each gave not only of themselves, but also pointed me in new directions, making connections with their colleagues and peers—each new relationship strengthened different theoretical threads and enriched the book's guiding narrative. To Patty Brennan, Michelle Fournet, Daniela Hedwig, Ron Hoy, Annie Lewandowski, Laura May-Collado, Bill McQuay, Rick Prum, Liz Rowland, and Peter Wrege, thank you. Above all, thanks and gratitude to Katy Payne, who sat down with me multiple times, read chapter drafts, and challenged me to think differently. Her way of being present with everyone and everything that surrounds animated this book from the beginning—without Payne, I am certain I never would have turned so fully to listening.

The project itself would not be as vibrant and interdisciplinary as it is without the incomparable Debra Hawhee. Thanks for giving me permission, time, and inspiration to play with sensory research methods, speculative theories, and more than academic writing. Thanks even more for reading draft after draft and fielding questions long after I graduated. The ideas and guidance of Rich Enos, Ann George, Cheryl Glenn, Mark Morrisson, Stuart Selber, Jack Selzer, and Brad Vivian also continue to inform all I do—gratitude for you all and to the programs at Texas Christian University and The Pennsylvania State University.

For the book's evolution from its early beginnings, I owe thanks to three anonymous reviewers and a series of readers and writing groups who engaged earlier drafts deeply, carefully, and rigorously. Thank you to reviewers for thinking with me, for unearthing the inconsistencies in my argument, for pushing me to stop avoiding aesthetics, for encouraging me that the revisions would be worth the effort. When I reached theory impasses, I turned to Casey Boyle, Jennifer Clary-Lemon, Daniel Gross, Diane Keeling, and Thomas Rickert. I'm grateful to be in a field with such brilliant, generous thinkers. If you pick almost any complex idea in this book, you can likely trace the nuances of that idea back to my conversations with one of them. Thanks, even more, to Kyle Wagner at the University of Chicago Press: No one understood my convictions for subverting the traditional academic manuscript quite like him. Moreover, the enthusiasm and careful attention of my book team—Erin DeWitt, Nathan Petrie, and Kristin Rawlings—has touched every page. Their patience with these interdisciplinary musings that diffract theory through narrative is unmatched.

To Tim Barr, Curry Kennedy, Heidi Nobles, Ismael Quiñones, Ashley Rae, Melissa Yang, and Miles Young, drinks at every conference from here until retirement should be on me: what you all saw in the earliest of drafts gave me the energy I needed to keep going. To Ben Firgens, Scott Gunter, Stephanie Kerschbaum, Annika Konrad, Tony Irizarry, Carrie Mott, Nikki Orth, Bryan

ACKNOWLEDGMENTS 145

Picciotto, Haley Schneider, and Neil Simpkins, thanks for the writing accountability, the existential questions about why we do research, the scholarly chats, the friendship, the Friday hype sessions. To independent editor Kimmie Farris, thanks for reading the rough drafts I was scared to show anyone else—your kindness and thoughtfulness were important, but your pointed questions and attention to detail were crucial for this book's development.

In my time as junior faculty at the University of Louisville, no one protected my time or supported my intellectual journey like my Department Chair Glynis Ridley. To Karen Chandler, Andrea Gaughan, Paul Griner, Lauren Heberle, Luz Huntington-Moskos, Tim Johnson, Frank Kelderman, Karen Kopelson, Deborah Lutz, Kristi Maxwell, Andrea Olinger, Susan Ryan, Mary P. Sheridan, Ian Stansel, Angela Storey, Sarah Strickley, Joe Turner, Bronwyn Williams, and so many more, thanks for being inspiring, supportive colleagues. And now, as I join a team of rhetoricians whose work helped pave the way for my own, I am grateful to extend the ideas here alongside Casey Boyle, Diane Davis, Rasha Diab, Linda Ferreria-Buckley, Scott Graham, Annie Hill, Jo Hsu, Mark Longaker, Jacqueline Rhodes, Donnie Johnson Sackey, and Clay Spinuzzi.

Along the way, this work was supported and enriched by fellowships and research funding at Penn State's Center for Humanities and Information and the University of Louisville's Commonwealth Center for the Humanities and Society, by research funding from the University of Louisville's Department of English, and by various workshops and institutes. This project originated in Diane Davis and Thomas Rickert's "Rhetoric and New Materialisms" Rhetoric Society of America (RSA) Institute seminar; grew in Byron Hawk, Diane Keeling, and Thomas Rickert's "The Futures of New Materialisms" RSA seminar; and coalesced in Ralph Cintron, Michelle Hall Kells, and Donnie Johnson Sackey's "Poieses of the Future: The Trandisciplinarity of Climate Change, Migration, and Land-Based Ethics" RSA seminar. Lawrence Revard read the pages related to his father's poetry more carefully than I could have ever imagined—I am deeply grateful for the care it took to attend so closely. Thanks as well to mentors like Nate Stormer, who talked with me about Susanne Langer; Scott Graham and Bridie McGreavy, who helped strategize how to balance book writing with grant work; and the group of rhetorical new materialisms scholars led by Joshua Hanan and Laurie Gries, who gathered to think about regenerative rhetorics.

Yet sometimes the most important support came from those who rarely asked about my book. Shavonnie Carthens and Eboni Neal Cochran, you changed what this profession means for me. Everyone on the Air Justice team, your energy, passion, commitment, and joy remind me why I get up in the mornings. Sarah Adams, Miriam Gonzales, Liz Kingham, Katie Kleinkopf, Layli Miron, Tyler Pool, Tony Reitz, Ariel Seay-Howard, and Layne Taylor, your friendship sustains me. Wes, you are closer than a brother and you fill my life with its most vivid colors. Stella and Eloise, watching you both grow is one of the greatest joys I have known—never stop being your brightest, fullest selves. Everyone in my family, no one makes me laugh or smile like y'all. My grandparents, no voices cheer as loudly or pray as fervently as yours do. Mom and Dad, every side job, every move, every phone call, every hug, I needed every one to get me here.

But more than anyone else, it was Andrew who endured all the hours, weekends, and absences involved in writing this book. Because you never asked it to be otherwise, never thought twice about my abandoning our home for a writing cabin in the woods, never yelled back when I was certain your noise in the other room was interfering with the drafting process, never asked me to be or work less, this book is for you.

Notes

Introduction

1. For more on the role of rhetoric in acts of witnessing that bring the past into the present, see Hawhee, *A Sense of Urgency*; Vivian, *Commonplace Witnessing*; and Boyle, "Aesthetics of Witnessing."

2. Studies of "nature" often generate from White settler scholars who idealize the land by overlooking how nature is interwoven with culture. In this schema, nature becomes a pristine ideal. Such ideals, rhetoric scholar Jennifer Clary-Lemon argues, recommit colonial harms. By contrast, scholars should "recognize the land-based politics of the field: who has cared for it over time; who currently lives on or occupies it; who returns to it, year after year; who changes it; and by what means; who is responsible for it; who tells its stories through their bodies" (Clary-Lemon, *Nestwork*, 13). My use of the term "natural environments" instead of "nature" is meant to acknowledge that nature is not a singular entity and to emphasize how interactions with environments are always situated through how an individual senses and experiences that environment. However, I retain "nature" when the thinkers I interview and engage with use that term. In chapter 1, for example, the erroneous divide between nature and culture is the issue under study for philosophers and rhetoricians of science in the second half of the twentieth century.

3. Notions of "kin" will feature most prominently in chapter 3—the term "nonhuman kin" is used to include more than nonhuman animals as communicative, agential beings in natural environments. A term that carries the idea of relation, "kin" has become an important concept for new materialist theories, feminist science studies, and the environmental humanities. Deborah Bird Rose relies on the idea of kin to point to how all beings, if not all elements, are connected and entangled (Van Dooren and Chrulew, *Kin*, 1). Donna Haraway likewise relies on "kin" as an "assembling sort of word" that extends beyond ancestry or genealogy (Haraway, *Staying with the Trouble*, 102–3). These notions of kin rely on the rich, historied valences of the term in Indigenous studies, and Robin Wall Kimmerer recently suggested an emphasis on "kinning" over "kin," so as to prioritize practices of making kin, rather than understanding kinship only theoretically (Van Horn, "Kinning," 7).

4. Whereas the ecological crisis in Oak Grove, Louisiana, serves as my personal tie to this study, climate change and environmental loss are featured only in chapter 4. For an extensive, complex study of the rhetorical role of emotion in our present climate crisis, see Joshua Trey Barnett's *Mourning in the Anthropocene*.

5. Animal rhetoric scholars have acknowledged nonhuman ways of knowing and communicating for some time. George Kennedy first directed attention to animal rhetorics in the 1980s, and since then Diane Davis, Carolyn Miller, Marilyn M. Cooper, and Louise Westling have shown how human language and agency cannot be separated from the nonhumans from whom we evolved (Kennedy, "A Hoot in the Dark"; Davis, "Some Reflections on the Limit"; Miller, "The Appeal(s) of Latour"; Cooper, *The Animal Who Writes*; Westling, *Logos of the Living World*). Kristian Bjørkdahl, Alex C. Parrish, Emily Plec, and Natasha Seegert have also analyzed how nonhuman kin employ rhetoric as well as determined possibilities for cross-species interaction (Bjørkdahl and Parrish, *Rhetorical Animals*; Parrish, *Sensory Modes of Animal Rhetorics*; Plec, *Human-Animal Communication*; Seegert, "Play of Sniffication"). My work is more interested in the rhetorical and semiotic work exchanged not only between humans and nonhuman animals but also amid the affective elements that emerge from natural environments in what Diane Keeling has called "a shared arena of sensation" (Keeling, "Feral Rhetoric").

6. Safina, *Becoming Wild*, 202.

7. Danto, *The Abuse of Beauty*, 7, 15, 27, 160.

8. Pippin, *After the Beautiful*, 13.

9. Pippin, *After the Beautiful*, 11.

10. Pippin, *After the Beautiful*, 135.

11. Hawhee, *Rhetoric in Tooth and Claw*, 5–6, 19. In *Rhetoric in Tooth and Claw*, Debra Hawhee charts the three valences of *aisthēsis* as found in Aristotle's rhetoric: the first, "feeling *aisthēsis*," is a way of "sensing pain and pleasure through touch and a host of other senses" that can be shared across species and is, in many ways, the kind of sensory encounter examined most closely throughout this book. "Deliberative *aisthēsis*" is more "*logos*-dependent perception," and "*phantasia*" is often associated with the imagination.

12. Mandoki, *Indispensable Excess of the Aesthetic*, 3–4.

13. Hawhee, *Rhetoric in Tooth and Claw*, 6.

14. Keeling, "Feral Rhetoric," 235.

15. Johnson confirms this linkage of aesthetics and rhetoric, as he defines aesthetics as the study of "visceral engagement with meaning" and how "meaning comes to us via patterns, images, concepts, qualities, emotions, and feelings that constitute the basis of our experience, thought, and language" (Johnson, *Aesthetics of Meaning and Thought*, 1).

16. McLeish, *Poetry and Music of Science*, 7.

17. Political scientist Susan Bickford supports this idea of the spectrum of rhetorical invention when she states, "What lies between representative thinking on the one hand, and a kind of impossible empathy on the other, is listening" (Gross, *Being-Moved*, 187n16). For more on rhetorical invention, see Atwill and Lauer, *Perspectives on Rhetorical Invention*; Muckelbauer, *The Future of Invention*; Pender, *Techne*; Rickert, "Toward the Chora"; and Simonson, "Reinventing Invention."

18. In contemporary rhetorical theory, new materialist approaches have been ushered in by studies of affect, materiality, and sensation. Rhetoric scholar Debra Hawhee argues for the importance of considering sensation as rhetorical when she argues that "sensation needn't become encased in language to be known" (Hawhee, "Rhetoric's Sensorium," 13). Ecological approaches to rhetoric have similarly allowed what were once thought of as discrete elements of communication—speaker/writer, audience, message, purpose, context—to bleed into one another; see Edbauer, "Unframing Models"; Clary-Lemon, *Nestwork*; Stormer and McGreavy, "Thinking Ecologically"; and McGreavy et al., *Tracing Rhetoric and Material Life*. Rhetorical

new materialisms via Thomas Rickert, S. Scott Graham, and Amy Propen even more fully observe boundary-marking practices amid the entanglement of the rhetorical, the material, the ecological, the posthuman, the ontological, and so forth (Rickert, *Ambient Rhetoric*; Graham, *Where's the Rhetoric?*; Propen, *Visualizing Posthuman Conservation*).

19. Maria E. Gigante, Jonathan Buehl, and Alan G. Gross and Joseph Harmon's work on image and multimodality in science as well as Gross's work on popular science writing serve as important exceptions (Gigante, *Introducing Science Through Images*; Buehl, *Assembling Arguments*; Gross and Harmon, *Science from Sight to Insight*; Gross, *The Scientific Sublime*). The three main approaches to rhetorics of science outlined by J. Blake Scott include: (1) close readings of texts by scientific heroes like Newton, Darwin, Watson, and Crick; (2) genre-discourse analysis that evaluates social and epistemic knowledge-making practices in scientific discourse communities; and (3) studies of rhetorical controversies (Scott, *Risky Rhetoric*). Landmark works in these approaches include Alan G. Gross, *The Rhetoric of Science*; Lawrence J. Prelli, *A Rhetoric of Science*; Jeanne Fahnestock, *Rhetorical Figures in Science*; Leah Ceccarelli, *Shaping Science with Rhetoric*; Carol Berkenkotter and Thomas N. Huckin, *Genre Knowledge in Disciplinary Communication*; Celeste Condit, *Meanings of the Gene*; and Jordynn Jack, *Autism and Gender*. Scott himself proposes a "rhetorical-cultural" approach, and scholars such as S. Scott Graham and Caroline Gottschalk Druschke have called for more interdisciplinary, collaborative work with scientists rather than studying science from a distance (Scott, *Risky Rhetoric*; Graham, "Promise and Peril of Scientific Science Studies"; Druschke et al., "Better Science Through Rhetoric").

20. Gross, *Being-Moved*, 13. Another touchstone with listening in rhetoric is Krista Ratcliffe's theory of "rhetorical listening," which shows that listening is not just a passive, receptive process, but one in which the listener is involved in processes of meaning making (Ratcliffe, *Rhetorical Listening*).

21. Keller, *Reflections on Gender and Science*; Harding, *Whose Science? Whose Knowledge?*; Haraway, *Modest_Witness*; Cipolla et al., *Queer Feminist Science Studies*; Booher and Jung, *Feminist Rhetorical Science Studies*; McKittrick, *Dear Science*.

22. What gets published in science remains caught up in racist, classist, and gendered norms; see Prescod-Weinstein, "Black Women Scientists Under White Empiricism."

23. Peters, *The Marvelous Clouds*, 91.

24. Both observation and listening are each multisensory, as scholars of multimodality show. Rhetoric scholar Steph Ceraso's contention is that "listening is a multisensory act" that should be conceived of as beyond the ear because listening "involves attending to not only the sensory, embodied experience of sound, but to the material and environmental aspects that comprise and shape one's embodied experience of sound" (Ceraso, "(Re)Educating the Senses," 102, 105).

25. Haraway's modest witness derives from Shapin and Schaeffer's account of scientific experimentation in *Leviathan and the Air-Pump*. Haraway diagnoses what has maintained the modest witness in science for so long as such: "Either critical scholars in antiracist, feminist cultural studies of science and technology have not been clear enough about racial formation, gender-in-the-making, the forging of class, and the discursive production of sexuality *through the constitutive practices of technoscience production themselves*, or the science studies scholars aren't reading or listening—or both" (Haraway, *Modest_Witness*, 35).

26. In "Situated Knowledges," Haraway famously described objectivity as "a conquering gaze from nowhere . . . that mythically inscribes all the marked bodies, that makes the unmarked category claim the power to see and not be seen, to represent while escaping representation" (581). In *Objectivity*, Lorraine Daston and Peter Galison, on the other hand, historicize objectivity and show

how it has been formed and enacted in different material ways with different scientific paradigms. Sandra Harding defines a "strong objectivity" as an objectivity that does not claim to be value-neutral but instead identifies and addresses researcher bias (Harding, *Objectivity and Diversity*, 32, 47). Karen Barad centers this idea of researcher bias as one of accountability, not only to one another but to the environment around us. "Objectivity," she argues, "is about being accountable to the specific materializations of which we are a part" (Barad, *Meeting the Universe Halfway*, 91).

27. Haraway, *Modest_Witness*, 25.
28. Haraway, *Modest_Witness*, 267.
29. Chuh, *The Difference Aesthetics Makes*, 22–23.
30. Chuh, *The Difference Aesthetics Makes*, 19.
31. Boyle, "Aesthetics of Witnessing," 46.
32. Peters, "Witnessing," 710.
33. Johnson, *Aesthetics of Meaning and Thought*, 1.
34. Keller, *A Feeling for the Organism*, 203.
35. McClintock quoted in Keller, *A Feeling for the Organism*, 203.

36. The question of "how" in scientists' ways of knowing animated this book from the beginning, and Boyle's work on the aesthetics of witnessing helped me articulate the "how" as a question of aesthesis. As he puts it: "The question is not . . . 'to be, or not to be,' but rather *how* to be and *how* not to be. It is the inflection of *how* that makes witnessing and bringing existence forth into the world an aesthetic problem" (Boyle, "Aesthetics of Witnessing," 57).

37. Any study of aesthesis that considers what cannot be contained by discourse or texts must consider theories of affect. Definitions of "affect" are often placed alongside definitions of "emotion." Brian Massumi creates a sharp distinction between emotion and affect: emotion is "subjective content" that derives from "the sociolinguistic fixing of the quality of an experience," and affect is an intensity that derives from a surfeit of "inexplicable," "unexpected," and "irreducible excess" (Massumi, *Parables for the Virtual*, 28). Brian Ott explains that Massumi's line of affect theory emerged in reaction to the "linguistic turn," which maintained that there is no "outside of the text" or no way to escape discursive language (Ott, "Affect in Critical Studies," 9). By contrast, Massumi argues that affect being in *excess* of the verbal does not signify, as representational, discursive language does. Rather than this excess furthering us from feeling, affect seems to mean or move more immediately. As Massumi has famously remarked, "The skin is faster than the word." If affect is about that which spills over language and defies neat categorization, then it is sensory excess (Massumi, *Parables for the Virtual*, 25). Ott further argues that Eve Kosofsky Sedgwick, in *Touching Feeling*, extends Damasio's line of affect theory in how she grounds affect in humans' "biological system" (Ott, "Affect," 8). Sedgwick offers tools for inducing "nondualistic thought," which she believes is the state in which individuals may access feeling and affect. Yet another line of affect theory studies how objects, environments, atmospheres, and orientations come to accrue human affective attachments—see Ahmed, "Happy Objects"; Berlant, "Cruel Optimism"; and Stewart, "Atmospheric Attunements." Finally, and most applicable to this study, in *The Ascent of Affect*, Ruth Leys analyzes how historical theories and scientific studies of emotion inform present theories of affect.

38. This mixed-methods approach also includes textual and contextual analysis of archival research. Archival collections consulted for this project include the Kenneth Burke Papers in the Eberly Family Special Collections Library at Penn State; materials obtained at the Burke house in Andover, New Jersey; the Susanne K. Langer Papers at the Houghton Library at Harvard University; as well as the György Kepes Papers at the MIT Special Collections Library. The intensive

interviews for this project were determined as IRB exempt and followed an informal structure that allowed for extemporaneous conversation and responsivity to how each subject wished to lead our conversation in divergent directions. Each subject, however, was asked at least five of the same questions concerning the feelings associated with their initial scientific findings in the field, what methods they employed to understand their subjects of study, how they defined objectivity, and what role beauty played (or not) in their work. Following Kathy Charmaz's *Constructing Grounded Theory*, I analyzed the interviews first with line-by-line coding and then coded for actions. I then engaged in comparative coding to determine which key terms emerged across all seven interviews; these keywords were *feeling, affect, beauty,* and *objectivity.*

39. Pink, *Doing Sensory Ethnography*, 74–76.

40. For more on the ethnographic turn in rhetoric studies, see Candice Rai and Caroline Gottschalk Druschke's *Field Rhetoric*.

41. Technical communication scholars Natasha Jones and Rebecca Walton draw an important distinction between story and narrative: whereas "stories are controlled by the teller . . . narratives are enacted through participation" by the people within and around the stories we tell. Because narrative promotes identification between individuals, encourages reflexivity, and interrogates historicity and context, it is a useful method for grappling with issues of social and racial justice (Jones and Walton, "Narratives to Foster Critical Thinking," 243). Jones also links narrative to listening when she argues that the sharing of stories requires a listener, which promotes contact across difference between listened to and listener (Jones, "Coalitional Learning").

Chapter One

1. There can be no underestimating the role of Burke in our ever-expanding definitions of rhetoric. As Debra Hawhee highlights in *Moving Bodies*, Burke's notions of rhetoric were transdisciplinary, drawing as much from mysticism as from biology. Yet as Diane Davis makes clear, Burke drew a staunch line between what is and is not rhetorical—rhetoric, he argued, had to do with symbolic thought, and only human beings were "symbol-using animals," by his account (Davis, "Some Reflections on the Limit").

2. No date is provided for Burke's return to reading Langer, but because *Philosophy in a New Key* was published in 1942 and Burke began reading it shortly thereafter only to return to the text much later, we can assume that a decade or more passed after the original publication and Burke's rereading of the text. Further, Burke writes a note at the bottom of his handwritten remarks that his analysis of Langer be "duplicated in FORM folder." Burke's most extensive remarks on form are found in *Rhetoric of Motives* and in the "On Form" essay published just before the "Definition of Man" in 1964.

3. Lewis A. Dexter, letter, August 1, 1962; Dryden, "Susanne K. Langer," 198; Howard Gardner, letter, July 21, 1970.

4. Langer, "Comments on Papers Read at the Conference of the American Academy of Arts and Sciences," May 6, 1956, Susanne K. Langer Papers. Houghton Library, Harvard University.

5. Rhetoric scholars may note major parallels between Burke and Langer. In striking ways, their respective careers appear as inverses of the other: Langer's attention moves from language early in her career to art later in her career, whereas Burke's attention moves from art to theories of language, as seen in his trajectory from *Counter-Statement* (1931) to *Language as Symbolic Action* (1966). Theories of mind and the biology of sensation maintained a steady presence in the work of both thinkers throughout the entirety of their careers.

6. Several scholars have advocated for the importance of Langer in rhetoric and composition studies. In the 1980s, James A. Berlin would cite Langer among a list of thinkers—including Burke, Richards, Cassirer, and Dewey—central to the emergence of "New Rhetoric," that school of thought that brought rhetoric back from its divorce from dialectic (Berlin, "Contemporary Composition," 773). Many years later, Arabella Lyon would reference Langer as the "mother of contemporary rhetorical theory" who tore symbols away from logic by theorizing how some symbols work in non-logical ways (Lyon, "Susanne K. Langer," 266, 269). Ann E. Berthoff also relies on Langer's theories to highlight the unnecessary dichotomy between thinking and feeling, which brings positivists to relegate imagination to the "affective domain" and logic to the "cognitive domain" (Berthoff, "The Intelligent Eye," 109).

7. Langer, *Mind*, 1:52.

8. Barthes, *Camera Lucida*, 51.

9. While philosophy was Langer's institutional home, her work has since found more advocates in music scholarship and art criticism. Only recently—now that some of her key concepts, such as vitality, are gaining traction with the emergence of new materialisms—has the field in which she trained come to respect her work. See Richerme, "A Deleuzian Reimagining."

10. First theorized by Donna Haraway as a counter to reflection, which disperses the same elsewhere, diffraction as a methodology was made more prominent by Karen Barad when she explained diffraction as a phenomenon that occurs when waves (light, sound, or water) overlap and combine, events that produce symmetrical "diffractive" patterns. Haraway, *Modest_Witness*; Barad, *Meeting the Universe Halfway*.

11. A first-generation immigrant from Germany, Langer grew up speaking German at home and began reading German philosophy, such as Kant's *Critique of Pure Reason*, as early as age twelve. While pursuing a philosophy degree at Radcliffe College, she balanced studies of formal logic with Henry M. Sheffer by learning music composition, mastering the cello, and penning a book of fairy tales. Sheffer's innovative approach to logic, which Alfred North Whitehead would cite as the greatest advance to modern logic, changed Langer's way of thinking, allowing her to "see logic as a field for invention"—a field that, although "traditionally stiff and scholastic," offered "as much scope for originality as in metaphysics" (Dryden, "Susanne K. Langer," 190–91).

12. Langer, *Feeling and Form*, 17.

13. Langer, *Feeling and Form*, vii.

14. Pippin, *After the Beautiful*, 13.

15. Langer, *Feeling and Form*, 394.

16. Langer, *Philosophy in a New Key*, 17. *Philosophy in a New Key* would sell over 500,000 copies, making it the most widely sold paperback edition of any scholarly manuscript for Harvard University Press in its time. The book was also one of the first originally written for an academic press that was then picked up by a mass-market publisher like Penguin. Schultz, *Cassirer and Langer on Myth*, 251; Dryden, "Susanne K. Langer," 195.

17. Langer, *Philosophy in a New Key*, 20.

18. Langer, *Philosophy in a New Key*, 21. In his own theory of symbolism, Whitehead distanced symbols from meaning by showing that there was a more "fundamental" symbolism apart from the symbolism in language and art. Sense perception, Whitehead argued, is a symbolic process, one whose meaning issues forth from nature rather than through social construction (Whitehead, *Symbolism*, 57).

19. Johnson's discussion of pragmatism aligns Langer with thinkers like Dewey, Peirce, and Russell. In rhetoric, John Muckelbauer is invested in rhetoric's asignifying work, to which Langer's theories would serve as a rich resource. Johnson, *Aesthetics of Meaning and Thought*, 219–23; Muckelbauer, "Asignification."

20. Contemporary evolutionary perspectives on music echo Langer's argument by showing the connections between music, communication, and nature. As neuroscientist Jay Schulkin contends, "Music is always centered on communication, imaginative or actual, and it is deeply rooted in nature. The core of music, just like the core of language, is calling out to others" (Schulkin, *Reflections on the Musical Mind*, 34).

21. Langer, *Philosophy in a New Key*, 233.

22. Langer, *Philosophy in a New Key*, 234.

23. Waldrop, "Inside Einstein's Love Affair with 'Lina.'"

24. Parr, "Barbara McClintock."

25. Gugleilmi, "Using Insect Eggs to Overturn Evolutionary Doctrine." For more on the role of music in the creative processes of famous scientists, see Tom McLeish's *The Poetry and Music of Science*.

26. Dryden, "Susanne K. Langer," 190–91.

27. Langer's notion of prescientific invention aligns with her contemporary Michael Polanyi's idea of the "tacit dimension," which reconsiders "human knowledge by starting from the fact that *we can know more than we can tell*" (Polanyi, *The Tacit Dimension*, 4).

28. Langer identified five idols of the laboratory: physicalism, jargon, methodology, objectivity, and mathematicization. These idols, she argued, ask scientists to identify problems and understand data only in accordance with the previously agreed-upon "approaches" within their fields of study, approaches inclusive of but not limited to the "technical language, the laboratory atmosphere, apparatus, graphs, charts and statistical averages" normative to their field (Langer, *Mind*, 1:33).

29. Whitehead, *Science and the Modern World*, 97.

30. Whitehead, *Science and the Modern World*, 199.

31. *Oxford English Dictionary*, "precious," accessed July 15, 2023, https://doi.org/10.1093/OED/1048217802.

32. Langer, *Mind*, 3:209.

33. Campt, *Listening to Images*, 6.

34. Campt, *Listening to Images*, 7.

35. Langer's theories of music and dance continue to resonate with contemporary scholars, but her writing is undercut at times by her adopting the perspective of early anthropological research that appropriates African music and communities in the Global South. Although Langer was invested in the earliest forms of human music to show the complexity of human behavior—not to elevate Western art music over these forms—this area of her work warrants critique and revision. Richerme, "A Deleuzian Reimagining"; Smolen, "Dance."

36. Danto, foreword to *Mind*, by Langer, v.

37. Within affect theory, there are important distinctions between feeling and emotion. Feeling is largely considered as emerging from nondiscursive elements or intensive force, and emotion as discursive articulations for those feelings. Langer's proposal that feeling is registered through nondiscursive or affective symbols keeps her ideas conversant with theories of affect.

38. Dryden, "Susanne K. Langer," 198.

39. Dryden, "Susanne K. Langer," 198.

40. Dryden, "Susanne K. Langer," 198.

41. As her teacher Whitehead would argue, abstraction is not only cerebral but natural: "Abstraction expresses nature's mode of interaction and is not merely mental. When it abstracts, thought is merely conforming to nature—or rather, it is exhibiting itself as an element in nature" (Whitehead, *Symbolism*, 26).

42. Langer explains how things (art objects or not) come to hold a sense of vitality through what she calls "the semblance of a thing," with "semblance" being defined as a thing's "direct aesthetic quality" (Langer, *Feeling and Form*, 50). Semblance is what gives art its strangeness, its ability to puncture or strike us. Affect theorist Brian Massumi defines Langer's "semblance" as something that occurs not only in art, but also in "natural perception." Semblance is not direct sense perception, as Massumi defines it, but a sort of awareness of perception: "This is a thinking of perception in perception, in the immediacy of its occurrence, as it is felt—a *thinking-feeling*" (Massumi, *Semblance and Event*, 44). Here, aesthesis is not abstract—an appreciation of a decorative motif because we are taught that this thing is beautiful—but directly emergent from the experience of feeling ourselves encounter said art. In this way, art gets through to the vitality subsistent within us and its semblance lingers, even if that semblance is never discursively known.

43. Langer's "felt life" could thus be considered an early precursor to feminist philosopher Elizabeth Grosz's "incorporeal" in that both concepts point theorists to the inarticulate something that animates matter, or brings the life of matter to our attention. To put it another way, life is an evolutionary process, and querying how life is felt through symbolic processes like art was one way to consider how humans can observe the evolutionary history within us. Our actions are enactments of evolutionary processes, and, according to Langer, the workings of art are one way to exemplify those processes, bringing us to feel all the life within us. Grosz, *The Incorporeal*.

44. Barthes, *Camera Lucida*, 51.

45. Barthes, *Camera Lucida*, 26.

46. Knuuttila, "*L'effet de réel* Revisited," 114.

47. Barthes, *Camera Lucida*, 42.

48. In his presentation of studium, Barthes positions himself as a "docile cultural subject," or a sort of modest witness, when he analyzes how the photograph "speaks." Barthes thus refuses to acknowledge, as Christina Sharpe exposes, how "race, place, class, aesthetics" and colonial logics influence ways of seeing. Supplementing Barthes's theory through Langer importantly highlights the individual subjectivity of sensing, but Sharpe's revision of the punctum is more important still: Sharpe's beautiful photographs of her Black mother disrupt Barthes's gaze, and though a punctive feeling from the photograph persists, this feeling promotes only a "deep regard" for Sharpe that returns the subject of the photograph to their opacity. Sharpe, *Ordinary Notes*, 178–79, 214.

49. Butchart, "Roland Barthes' *Camera Lucida*," 213, 200.

50. The etymology of "punctum" furthers the idea of a stop in interpretation, as one valence of the term is that of a punctuation mark indicating a "full stop." Spatially, the punctum is a geometric point. Zoologically, the punctum is a rounded spot of color. Temporally, the punctum marks a brief moment, an instant. Across all definitions, the punctum marks something that stops you in your tracks. What is unique for Barthes is the layer of affect that lingers in the attention beyond the moment where a viewer turns away. *Oxford English Dictionary*, "punctum," accessed July 15, 2023, https://doi.org/10.1093/OED/6783302995.

51. Similar to how Whitehead gives nature autonomous power to make meaning for symbols, historian Eelco Runia argues that Barthes grants the punctum autonomous power that "exists independently of its receptivity" (Runia, *Moved by the Past*, 100). As Runia further shows, the punctum works metonymically by bringing what is *not* present into the scene. Metonymy, Runia goes on to explain, is "not an exclusively linguistic phenomenon" and can work in other mediums, like photography (Runia, "Presence," 16). Just as metonymy is not exclusively linguistic, the punctum is not exclusively visual.

52. Langer, *Mind*, 1:20.

53. Langer, *Mind*, 1:57.

54. Langer to György Kepes, December 31, 1958, Susanne K. Langer Papers. Houghton Library, Harvard University.

55. Fricker, "Intuition and Reason," 184. Fricker also shows the gender disparity of intuition in "Why '*Female* Intuition'?"

56. Langer, "The Great Shift."

57. Langer, "Artistic Perception," 7; emphasis in original.

58. In one sense, Langer's intuition functions like Burke's "piety": a sense of what properly goes with what, an automatic feeling or interpretation of forms. Intuition is the capacity for making sense of what happens by bringing all elements of experience together for a resonance of sorts. Burke, *Permanence and Change*.

59. Langer, "The Mind," 10–11.

60. Langer, "The Mind," 11.

61. This idea of a subject/object distinction that occurs during the rhetorical encounter foreshadows Barad's agential realism in *Meeting the Universe Halfway*.

62. Nocek, "A Feeling for Biological Concepts," 266. In how her biology was shaped by her aesthetics, Nocek argues, Langer's biological theories align with early theories of epigenetics.

63. Langer, *Feeling and Form*, ix.

64. Langer, *Mind*, abridged ed., 104.

Chapter Two

1. Emphasis added.

2. According to editors John Bryant and Haskell Springer, "humane" was altered to "human" in the British edition of *Moby-Dick* (Melville, *Moby-Dick*, 601). It is unclear, though, whether this difference was the result of a typographical error or whether Melville intended "humane" to invoke the idea of humanity facing natural terror.

3. This chapter is derived in part from the article "Witnessing the Open Semiosis: A Method for Rhetorical Listening Beyond the Human," *Rhetoric Society Quarterly* 53, no. 1 (2023): 30–44, © The Rhetoric Society of America, available online at https://www.tandfonline.com/doi/full/10.1080/02773945.2022.2078870.

4. To listen to the rings of Saturn, visit "NASA Voyager Space Sounds," https://www.youtube.com/watch?v=at3RXrUwMGg.

5. To listen to these songs, visit "Songs of the Humpback Whale," https://www.youtube.com/watch?v=sjkxUA041nM.

6. R. Payne, *Among Whales*, 166.

7. R. Payne and McVay, "Songs of the Humpback Whale," 585; emphasis added.

8. R. Payne, *Among Whales*.

9. K. Payne, interview by author, 3.

10. Hereafter, in the text I will use "Payne" to reference Katy Payne and use Roger Payne's full name to avoid confusion with their shared last name.

11. K. Payne, "Progressively Changing Songs."

12. Rothenberg, *Thousand Mile Song*, 45.

13. Rothenberg, *Thousand Mile Song*, 23–25.

14. Rothenberg, *Thousand Mile Song*, 3.

15. K. Payne, interview, 2.

16. Melville, *Moby-Dick*, 404.

17. K. Payne, Interview by author, 2.

18. The songs featured in *Science* were based on the recordings of naval engineer Frank Watlington, who was the first to take the Paynes out in the Atlantic Ocean to hear the songs of the humpback whale. Later, the Paynes would go on to conduct their own recordings of humpback whales off the coast of Hawaii and of right whales off the coast of Patagonia.

19. R. Payne and McVay, "Songs of the Humpback Whale," 585.

20. I should note that Roger Payne and McVay's definition of these sounds as "songs" follows from biologists' definition of birdsong, the main characteristic of which is that song follows fixed patterns of sound. Interestingly enough, when the researchers sped up the whale recordings, which subsist of elongated sounds, the whale songs followed patterns and sounds similar to birdsong. For a detailed explanation of why some animal sounds can be considered songs, see violinist, composer, and ornithologist Hollis Taylor's *Is Birdsong Music?*

21. R. Payne, *Among Whales*, 22.

22. K. Payne, interview by author, 3.

23. K. Payne, interview by author, 6.

24. R. Payne, *Among Whales*, 166.

25. Rothenberg, *Thousand Mile Song*, 30.

26. By "strangeness" here, I don't mean to imply dissonance. This was not confusion or discomfort. The strangeness stemmed from finding the songs familiar, almost commonplace, but not being able to articulate what was commonplace or understand why I felt that way.

27. K. Payne, interview by author, 3–4.

28. K. Payne, interview by author, 12.

29. Nezhukumatathil, *World of Wonders*, 156.

30. K. Payne, "Elephants and Whales," 18.

31. R. Payne and McVay, "Songs of the Humpback Whale," 590.

32. K. Payne, "Progressively Changing Songs."

33. R. Payne, *Among Whales*, 156.

34. K. Payne, "Progressively Changing Songs," 142.

35. Whitehead and Rendell, *Cultural Lives of Whales and Dolphins*, 91.

36. Whitehead, "Society and Culture in the Deep and Open Ocean," 456.

37. Xanthopoulos, *Fathom*; Garland et al., "Transmission of Humpback Whale Song."

38. Herman, "Multiple Functions of Male Song," 1798.

39. Qtd. in McQuay and Joyce, "It Took a Musician's Ear."

40. Herman, "Multiple Functions of Male Song," 1801.

41. K. Payne qtd. in Taylor and Flatow, "How the Humpback Says Hello."

42. Whitehead and Rendell, *Cultural Lives of Whales and Dolphins*, 82.

43. K. Payne, interview by author, 15, 13.

44. K. Payne, interview by author, 15.
45. K. Payne, interview by author, 2.
46. K. Payne, qtd. in Brody, "Scientist at Work," 2.
47. K. Payne, interview by author, 27.
48. There is much to suggest that the overlap of theory and practice between Payne and Langer has to do with a shared reverence for nature. At least that was the case for Langer's teacher Whitehead, who considered the role of the observer in nature as more of a witness. Stengers interprets as much when she translates Whitehead's ideas as follows: "We know that if we pay [nature] due attention, we will find more in it than what we observe at first glance. The sense organs testify to the importance of paying due attention to nature ... and the body is therefore not what explains but what testifies" (Stengers, *Thinking with Whitehead*, 69).
49. K. Payne, interview by author, 29.
50. McQuay, interview by author, 15.
51. Qtd. in McQuay and Joyce, "It Took a Musician's Ear," para. 9.
52. Qtd. in McQuay and Joyce, "It Took a Musician's Ear," para. 11.
53. K. Payne, interview by author, 19.
54. K. Payne, interview by author, 2.
55. K. Payne, interview by author, 7; emphasis added.
56. Kohn, *How Forests Think*, 30.
57. Kohn, *How Forests Think*, 31.
58. Kohn, *How Forests Think*, 66–68.
59. In *The Sensory Modes of Animal Rhetorics*, Parrish explains via Peirce's "sign properties" how this detached perspective arises: "Firstness is simply a sign's feeling or one's sense of a sign. Secondness is the level of physical fact, of a sign's material reality. Thirdness is the level of general rules that governs firstness and secondness in any given object." To symbolize, then, is to be caught up in thirdness, or to be able to consider how the symbol functions via cultural influence. Parrish, *Sensory Modes of Animal Rhetorics*, 116.
60. Kohn, *How Forests Think*, 43.
61. Kohn, *How Forests Think*, 48.
62. Davis, "Some Reflections on the Limit," 278, 280.
63. Parrish, *Sensory Modes of Animal Rhetorics*, 24.
64. Kohn, *How Forests Think*, 9.
65. Kohn, *How Forests Think*, 33.
66. Kohn, *How Forests Think*, 34.
67. Goodale, *Rhetorical Invention of Man*, 4.
68. Kohn, *How Forests Think*, 42.
69. Kohn, *How Forests Think*, 72. The field of biosemiotics studies this open sharing of signs between human animals and the natural world, even considering how signals are sent within the human body. Jesper Hoffmeyer works parallel to Kohn when he posits that human animals are able to signify about the natural world because the natural world is itself signifying. "How can signification arise out of something that signifies nothing?" Hoffmeyer asks. Hoffmeyer, too, like Rickert, relies upon Jacob von Uexküll's theory of *Umwelt* to theorize communication and meaning beyond the human. For Hoffmeyer and others in biosemiotics, *Umwelt* comes to explain how all organisms live first and foremost in their own unique "semiospheres" (Hoffmeyer, *Biosemiotics*, vii, 3). Parrish further highlights that zoosemiotics also treats the sign as the basic unit of life. Kohn thus aligns with these arguments but would perhaps avoid the bio- and

zoo- distinctions, as, for him, semiosis is an open whole (Parrish, *Sensory Modes of Animal Rhetorics*, 44).

70. Kohn, *How Forests Think*, 222.
71. Eidsheim, *Sensing Sound*, 158.
72. Whereas this work is invested in considering the punctum as an important term for attention, Barthes specifically considers music interpretation in "The Grain of the Voice." Most consider "the grain" as the influence of the body in a singer's voice, but Barthes is also considering the immaterial elements of sound, given his turn to Julia Kristeva to distinguish between *pheno-song*, having to do with the expression and representation of song, and *geno-song*, having to do with the volume of the singing voice. Barthes, "Grain of the Voice," 182; Kristeva, *Revolution in Poetic Language*.
73. Eidsheim, *Sensing Sound*, 54.
74. K. Payne, interview by author, 24.
75. Guinee and K. Payne, "Rhyme-like Repetitions in Songs."
76. K. Payne, interview by author, 25.
77. K. Payne, interview by author, 22.
78. *Oxford English Dictionary*, "comportment," accessed July 15, 2024, https://doi.org/10.1093/OED/6304636789.
79. K. Payne, interview by author, 19.
80. McQuay, interview by author, 7.
81. McQuay, interview by author, 16.
82. K. Payne, interview by author, 20.
83. Oliver, "Witnessing, Recognition," 475.
84. Oliver, "Witnessing, Recognition," 476.
85. Sakakibara, *Whale Snow*, 19–20.
86. K. Payne, interview by author, 23.
87. K. Payne, "Elephants and Whales," 10.
88. K. Payne, interview by author, 17.
89. K. Payne, interview by author, 17.
90. K. Payne, interview by author, 18.
91. Qtd. in Nutt, "Creativity of Whale Songs," 3.
92. Haraway, "Situated Knowledges," 581.
93. K. Payne, interview by author, 18.
94. K. Payne, interview by author, 21.
95. K. Payne, interview by author, 7.
96. We should also consider the limits of Payne's praxis, requiring as it does a certain level of access, privilege, and ability not equitably available. Payne herself admits that "nobody told me I was going to have to earn my living through what I did in college." And it was her continual access to Ivy League resources that has enabled her to spend years with her material without needing to produce something from it. For so many women in science—especially women of color, individuals who experience disabilities, and other individuals who have been historically marginalized—that is not the case, as the other case studies in this book will show. Payne's objectivity, then, certainly has feminist implications but requires further theorizing through an intersectional lens. K. Payne, "Elephants and Whales," 2.
97. Hoy, Interview, 38.
98. Limón, "The Whale and the Waltz Inside of It," lines 42–54.

Chapter Three

1. For literary analysis of the title poem of the collection "How the Songs Come Down," see Lee's "Survivance Memories."

2. In the full poem, Revard calls upon the ears and understanding of formerly enslaved peoples who turned threatening sounds like "ambulance sirens" into the blues songs of resistance and protest. In their discussions of the poem and this theme across Revard's work, Lawrence Revard relays how his father considered song as a matter of genesis and survival (Revard, email message to author, November 24, 2024). An extension of this chapter would engage the implications of "Songs of the Wine-Throated Hummingbird" more deeply, considering not just the impossibility of fixing meaning in language but also how song amplifies agency, opening the creative powers of refusal and resistance for those under threat.

3. The term "dominant science" follows the work of Max Liboiron in *Pollution Is Colonialism*.

4. Weilgart, Whitehead, and K. Payne, "A Colossal Convergence." Melville also compares whales with elephants in chapter 86, of *Moby-Dick*, "The Tail."

5. To listen to soundscapes in the African rain forest, visit McQuay, "Night at Dzanga Bai Clearing," https://soundcloud.com/npr/night-at-dzanga-bai-clearing.

6. K. Payne, *Silent Thunder*, 15; K. Payne, "Elephants and Whales," 4.

7. K. Payne, "Elephants and Whales," 4.

8. K. Payne, Langbauer, and Thomas, "Infrasonic Calls," 297.

9. Elephant Listening Project, "About Us."

10. Payne outlines the three remaining elephant species and their respective regions as such: African savanna elephants live in southern and East Africa savannas with one remaining population in western Namibia; African forest elephants live in West and Central African equatorial rain forests; Asian elephants live in the forests and forest edges of India as well as across continental Southeast Asia and island Asia. K. Payne, "Three Elephant Species," 57.

11. K. Payne, Langbauer, and Thomas, "Infrasonic Calls," 297, 298.

12. Ben-Ari, "Throbbing in the Air," 354. There may also be visual evidence of infrasonic rumbling: K. Payne, Langbauer, and Thomas comment on a "fluttering, with an excursion of about 1 cm, of the area on the elephant's forehead where the nasal passage enters the skull" ("Infrasonic Calls," 298). This report has not been confirmed—or commented on, by my reading—from other researchers.

13. K. Payne, Langbauer, and Thomas, "Infrasonic Calls," 300.

14. Poole et al., "Low Frequency Calls of African Elephants," 392.

15. K. Payne, "Three Elephant Species," 62.

16. Safina, *Beyond Words*, 78–79, 92.

17. Safina, *Beyond Words*, 39.

18. Hoffman, "A Language of Listening," 53.

19. Wrege, interview by author, 8.

20. The idea of "significance" plays a key role across our texts of interest. See Barthes, "Grain of the Voice"; Kristeva, *Revolution in Poetic Language*; Langer, *Philosophy in a New Key*.

21. K. Payne, *Silent Thunder*, 28.

22. K. Payne, interview by author, 22.

23. Boyle, "Aesthetics of Witnessing," 45–46.

24. Revard, "Songs of the Wine-Throated Hummingbird," in *How the Songs Come Down*.

25. Robinson, *Hungry Listening*, 71; emphasis added.

26. Robinson, *Hungry Listening*, 50, 51.
27. Robinson, *Hungry Listening*, 45.
28. Liboiron, *Pollution Is Colonialism*, 6–7.
29. Tuhiwai Smith, *Decolonizing Methodologies*, 1.
30. Tuhiwai Smith, *Decolonizing Methodologies*, 61.
31. Robinson, *Hungry Listening*, 53.
32. Robinson, *Hungry Listening*, 71–72.
33. Robinson, *Hungry Listening*, 15; emphasis added.
34. TallBear, "Standing with and Speaking as Faith."
35. Robinson, *Hungry Listening*, 244.
36. Oliveros, *Deep Listening*, xxiii.
37. To put it another way, the punctum is more than the visual, the sonic is more than the aural, and the environment disrupts each mode, intermixing the aesthetics of multiple species. There is no passive scene to be sensed but, instead, an ongoing interplay of aesthetic occurrences.
38. Kimmerer, *Braiding Sweetgrass*, 48–59; Van Horn, "Kinning," 7.
39. Van Horn, "Kinning," 6.
40. Robinson, *Hungry Listening*, 15.
41. Kimmerer, *Braiding Sweetgrass*, 48–49.
42. Robinson, *Hungry Listening*, 53.
43. Robinson, *Hungry Listening*, 52.
44. Global Conservation, "African Forest Elephant."
45. Larson, "History of the Ivory Trade."
46. World Wildlife Fund, "Status of African Elephants." Wrege et al. report on the effects of seismic prospecting for oil on African forest elephants in "Use of Acoustic Tools."
47. Rowland, interview by author, 21.
48. Rowland, interview by author, 24–25.
49. Wrege, interview by author, 12.
50. Emlen and Wrege, "White-Fronted Bee-Eaters."
51. Emlen, Wrege, and Webster, "Cuckoldry as a Cost of Polyandry."
52. Wrege, interview by author, 5.
53. Wrege, interview by author, 16.
54. Safina, *Beyond Words*, 105.
55. Wrege, interview by author, 16.
56. Runia, "Presence," 1, 5.
57. Safina, *Beyond Words*, 83.
58. Wrege, interview by author, 9.
59. Wrege et al., "Acoustic Monitoring for Conservation," 1292.
60. Rowland, interview by author, 10.
61. Wrege, interview by author, 10.
62. Wrege, interview by author, 10.
63. Speaking of training from Western ways of knowing, artificial intelligence (AI) detectors that analyze data from the acoustic monitoring grid have proved no better, at least as of early 2024 when this book moved to publication. For a while, the Elephant Listening Project held out hope that AI could offer real-time acoustic monitoring, resulting in less time hand-browsing data and more time researching and working in the field. Not only is the technology still too

NOTES TO PAGES 75–80

expensive, but a human data analyst must sift through the faint rumbles missed by AI. Gomes-Selman et al., "Curriculum Learning."

64. Swider et al., "Passive Acoustic Monitoring."

65. In one report of Payne's work on elephants' infrasonic calls, Payne asks the researcher to attend "not only [to] the timing of calling" but also to the timing of "'elephants' bouts of listening' to determine whether elephants display more listening behavior at times when sound transmission is best" (Ben-Ari, "Throbbing in the Air," 357).

66. Safina, *Beyond Words*, 107.

67. *Oxford English Dictionary*, "resonance," accessed July 15, 2023, https://doi.org/10.1093/OED/9474612525.

68. Nancy, *Listening*, 3.

69. Nancy, *Listening*, 6.

70. For the role of resonance in rhetoric, see Byron Hawk's "Sound: Resonance as Rhetorical."

71. Van Dooren and Chrulew, *Kin*, 1.

72. Nancy, *Listening*, 10.

73. Communication and media scholar John Durham Peters complements this idea when he differentiates the visual from the aural through Peirce's icon and symbol. Vision, Peters argues, functions iconically, whereas sound functions symbolically, requiring associations that cannot be known intuitively or naturally, but associations that must be shared, or through a meaning-making process that requires ongoing social participation with others. As Peters puts it, hearing "is basically 'symbolic' in that it grasps the world through learned habit or convention" (Peters, *The Marvelous Clouds*, 303). These marked differences between the visual and the aural are precisely what lends Barthes's "punctum" such flexibility. By *not* being part of what is iconic in the visual, the punctum persists in meaning symbolically.

74. Ethnomusicologist and Colombian musician Ana María Ochoa Gautier's *Aurality* provides a detailed study of how listening is central to politics for living, precisely because of how sound implies "an ontology of relation." As Gautier puts it, "Entities that listen and entities that produce sound are entangled in the relation between nature and culture and mutually produce each other" (Gautier, *Aurality*, 22).

75. Langer, *Philosophy in a New Key*, 93.

76. Hedwig, interview by author, 8.

77. Hedwig, interview by author, 8–9, 4.

78. Hedwig, interview by author, 4.

79. Wrege, interview by author, 19.

80. Wrege, interview by author, 19–20.

81. In Kenneth Burke's house, where bookshelves were organized thematically, the work of Susanne Langer sat on a bookshelf directly next to Konrad Lorenz's *King Solomon's Ring*.

82. Safina, *Beyond Words*, 26–27.

83. Safina, *Beyond Words*, 27.

84. Williams, "Four Turtles," 65, 66.

85. Safina, *Beyond Words*, 35.

86. Hedwig, interview by author, 20.

87. Hedwig, interview by author, 20.

88. Hedwig, interview by author, 11.

89. Kimmerer, *Braiding Sweetgrass*, 208.

90. Wrege, interview by author, 33.
91. Wrege, interview by author, 29, 30, 31.
92. Wrege, interview by author, 24.
93. Wrege, interview by author, 22.

Chapter Four

1. Moore, *We Are All Whalers*.
2. Lewandowski, McDonald, and Rubin, *Siren*.
3. Lewandowski, interview by author, 29, 17.
4. Lewandowski, interview by author, 25.
5. Lewandowski, interview by author, 10, 19.
6. Kimmerer, "Attention, Curiosity, Play, Gratitude," 142.
7. Rivers, "Better Footprints," 174.
8. For a detailed account of how soundscapes have changed over time, as well as how different species interact in given soundscapes, see R. Murray Schafer's *The Soundscape*.
9. May-Collado, interview by author, 2.
10. Lopez, "Night Chirps in Costa Rica."
11. May-Collado, interview by author, 2.
12. May-Collado, interview by author, 3, 4.
13. May-Collado, interview by author, 3.
14. May-Collado, interview by author, 5.
15. Questions derive from May-Collado, Agnarsson, and Wartzok, "Body Size and Tonal Signals Frequency"; May-Collado, Agnarsson, and Wartzok, "Tonal Sound Production in Whales"; May-Collado and Wartzok, "Comparison of Bottlenose Dolphin Whistles"; and May-Collado, "Changes in Whistle Structure."
16. May-Collado, interview by author, 6.
17. May-Collado, interview by author, 6, 6–7. May-Collado's final conclusion on the river dolphins is that they use whistles but in a drastically different way than bottlenose dolphins. Rather than for "individual identity" or communicating to identify other animals in family groups or clans, river dolphins "use [whistles] to keep each other apart" and keep populations discrete. As a result, they don't produce whistles very frequently.
18. Sitar et al., "Tourists' Perspectives on Dolphin Watching," 79.
19. Kassamali-Fox et al., "Tour Boats Affect the Activity Patterns."
20. Lewandowski, interview by Wong, 6.
21. Perez-Ortega et al., "Dolphin-Watching Boats Affect Whistle Frequency."
22. Gagne et al., "Dolphin Communication," 1.
23. May-Collado, interview by author, 10.
24. May-Collado, interview by author, 12.
25. De Brabandere, "What Do You Hear Underwater?," para. 3.
26. Quoted in May-Collado, http://www.lauramay-collado.com/about-me.html.
27. May-Collado, interview by author, 20.
28. Lindsey and Dahlman, "Global Temperature"; Thompson, "Hottest Month Ever Recorded."
29. Cornwall, "In Hot Water," 445.
30. NASA, "How Long Have Sea Levels Been Rising?"; NOAA, "Sea Level Rise by 2050."
31. Cornwall, "In Hot Water," 442.

32. Cornwall, "In Hot Water," 442–43.
33. Elliott and Simmonds, "Whales in Hot Water?," 3.
34. NOAA Fisheries, "Whales and Climate Change."
35. Berkelhammer, "There Are Signals."
36. Okun, "Sense of Urgency."
37. When it comes to climate justice, those who experience the worst environmental impacts from rising sea levels and temperatures are those who often contributed least to climate change.
38. Hawhee, *A Sense of Urgency*, 9.
39. Vivian, *Commonplace Witnessing*, 4.
40. Peters, "Witnessing," 708.
41. *Oxford English Dictionary*, "emergency," accessed July 15, 2023, https://doi.org/10.1093/OED/8629629926.
42. Ayto, *Dictionary of Word Origins*, 57.
43. Bedau and Humphreys, *Emergence*, 1.
44. Bar-Yam, "Emergence," para. 1. Emergence also works fractally, as physicist Yaneer Bar-Yam defines it: "Emergence refers to how collective properties arise from the properties of parts, how behavior at a larger scale arises from the detailed structure, behavior and relationships at a finer scale" (Bar-Yam, "Scale," para. 5).
45. brown, *Emergent Strategy*, 3.
46. brown, *Emergent Strategy*, 13.
47. Bar-Yam, "Emergence," para. 3.
48. brown, *Emergent Strategy*, 13. For more Black feminist connections to marine mammals, see Alexis Pauline Gumbs's *Undrowned* and Micheala Harrison's *Whale Whispering* blog.
49. Quoted in brown, *Emergent Strategy*, 14.
50. "Urgency" comes from a different Latin root—rather than *mergere*, "urgency" is borrowed from the Latin *urgere*, which means to "push, press, compel," thus the idea of pressing forward or a pressing matter. The modern idea of emergency, then, is not just when something issues forth, but when something *urgent* issues forth, something pressing that disrupts our attention. Ayto, *Dictionary of Word Origins*; Partridge, *Origins*, 811; de Vaan, *Etymological Dictionary of Latin*, 259.
51. Fournet, interview by author, 6.
52. Fournet, interview by author, 6.
53. Fournet, interview by author, 19.
54. Fournet, interview by author, 4.
55. For a list of the sixteen call types of the humpback whale, see Fournet, Szabo, and Mellinger's "Repertoire and Classification of Non-Song Calls."
56. Xanthopoulos, *Fathom*.
57. Fournet, interview by author, 11.
58. Fournet, interview by author, 11.
59. Nancy, *Listening*, 31.
60. Fournet, interview by author, 12.
61. Fournet et al., "Some Things Never Change."
62. Fournet, interview by author, 3.
63. Fournet, Matthews, and Gabriele, "Impact of the 'Anthropause.'"
64. May-Collado, interview by author, 1.
65. May-Collado, interview by author, 15.

66. May-Collado, interview by author, 15.

67. May-Collado, interview by author, 15–16.

68. May-Collado, interview by author, 18, 16. May-Collado here conveys an emotional experience of discomfort rather than a bodily experience of hurt when she discusses pain. However, as rhetoric scholar V. Jo Hsu discusses in *Constellating Home*, understanding social perceptions of pain through the lens of disability studies would be an important extension of this study. Disability, too, surrounds chapter 5 and the work of Richard O. Prum, who moved to study the coloration of birds when a virus he contracted during fieldwork resulted in a major hearing loss that prevented him from detecting birdcalls and birdsongs.

69. May-Collado, interview by author, 18.

70. May-Collado, interview by author, 18.

71. May-Collado, interview by author, 15.

72. May-Collado, interview by author, 15.

73. In Kimmerer's experience, "when we put scientific instruments in their hands, they trust their senses less" (Kimmerer, *Braiding Sweetgrass*, 224).

74. Fournet, interview by author, 21.

75. Fournet, interview by author, 22.

76. Fournet, interview by author, 21.

77. Fournet, interview by author, 21–22.

78. Fournet, interview by author, 22.

79. *Oxford English Dictionary*, "siren," accessed July 15, 2023, https://doi.org/10.1093/OED/8379975644.

80. To listen to this bell-tower performance, visit "Annie Lewandowski | Cetus: Life After Life," https://www.youtube.com/watch?v=I76ure6WCSs.

81. Fournet, interview by author, 22.

82. Fournet, interview by author, 23.

83. Fournet, interview by author, 23.

84. Fournet, interview by author, 10–11.

85. Fournet, interview by author, 11.

86. Fournet, interview by author, 19.

87. Kimmerer, *Braiding Sweetgrass*, 300.

88. Kohn, *How Forests Think*, 72; Peters, *The Marvelous Clouds*, 382.

Chapter Five

1. Prum, "Re: Interview," email to author, September 3, 2019.

2. Prum, *Evolution of Beauty*, 345.

3. Revard, *How the Songs Come Down*, i.

4. Prum and Brush, "Which Came First?"; Prum, "Evolution of Structural Color," 4.

5. To see the dance of the manakins, watch Cordey's *Dancing with the Birds*.

6. Prum, *Evolution of Beauty*, 8.

7. Prum, *Evolution of Beauty*, 8.

8. To watch these wing songs, visit "Club-Winged Manakin Dance," https://www.youtube.com/watch?v=tSHjhCN6NC0.

9. Prum, *Evolution of Beauty*, 131.

10. Prum, *Evolution of Beauty*, 131; Prum, qtd. in Radiolab, "The Beauty Puzzle."

11. Prum, *Evolution of Beauty*, 73.
12. Coyne, "Prum's Book on Sexual Selection," para. 4.
13. Coyne, "Which Book Would You Hurl?," para. 4.
14. Borgia and Ball, review of *The Evolution of Beauty*, 188. Coyne further regards Prum's feminist epistemology in *Evolution* as "a particularly invidious way to sell his theory, as female choice in birds is a direct product of evolution, while human feminism is a rational conclusion our species draws to improve society by treating people equally" (Coyne, "Prum's Book on Sexual Selection"). Prum's response to Coyne, Borgia, Ball, and others is simply that "they're so pissed off they can't even read [the book]," much less engage with the arguments therein (Prum, interview by author, 20).
15. Prum, *Evolution of Beauty*, 177–78.
16. In her history of the naturalistic fallacy, feminist science studies scholar Lorraine Daston articulates that all "versions of the naturalistic fallacy [have] less to do with nature than with a militantly policed border between whatever two realms are put asunder," in this case nature versus culture (Daston, "Naturalistic Fallacy Is Modern," 586).
17. Prum, interview by author, 9.
18. Patricelli, Hebets, and Mendelson, review of *The Evolution of Beauty*, 122.
19. Prum, "Lande-Kirkpatrick Mechanism."
20. Prum, *Evolution of Beauty*, 40.
21. Levine, "Why Beauty Matters," 250; emphasis added.
22. Levine, "Why Beauty Matters," 262.
23. Bostwick, qtd. in Radiolab, "The Beauty Puzzle."
24. Prum, interview by author, 13.
25. Amundson, *Changing Role of the Embryo*, 8.
26. Prum, interview by author, 13, 12–13.
27. Prum, "Re: Interview," email to author, September 3, 2019.
28. Amundson, *Changing Role of the Embryo*, 1.
29. In the previous chapter, we learned that May-Collado finds placing ontogenetics and phylogenetics into tension with one another essential, like the balance between micro and macro involved in studies of emergence.
30. That Amundson's thesis was unsuccessful in changing scientists' approach to the hegemonic adaptation narrative can be seen in how infrequently evolutionary biologists cited his work. Throughout his career, Amundson was cited less than 2,500 times, and *The Changing Role of the Embryo in Evolutionary Thought*, the text that had so much influence on Prum, was cited less than 500 times. By comparison, Prum—in the middle of his career—has already accrued over 10,000 citations.
31. Prum, interview by author, 13–14.
32. Prum, interview by author, 13.
33. Prum, interview by author, 13.
34. Brennan, interview by author, 19.
35. Prum, interview by author, 18.
36. This evolutionary defense mechanism ensures that although nearly 40% of total copulation in several duck species occurs via forced copulation, only 2–5% of offspring are fathered by a male who was not the chosen partner of the female. Prum, *Evolution of Beauty*, 172–73.
37. Prum, interview by author, 18, 19.
38. Prum, interview by author, 19.

39. Brennan and Prum, "Limits of Sexual Conflict."
40. Brennan, interview by author, 3.
41. Brennan, qtd. in Radiolab, "The Beauty Puzzle."
42. Brennan, interview by author, 4.
43. Brennan, interview by author, 19.
44. Brennan, interview by author, 18.
45. Brennan, interview by author, 8–9.
46. Brennan, interview by author, 2.
47. Brennan, interview by author, 2.
48. Brennan, interview by author, 2.
49. Prum, "Duck Sex," para. 1.
50. Brennan, "Why I Study Duck Genitalia," para. 1.
51. Brennan, "Why I Study Duck Genitalia" para. 8; emphasis added.
52. Brennan, interview by author, 9.
53. Prum, *Evolution of Beauty*, 143–44.
54. Hawhee, *Rhetoric in Tooth and Claw*, 13–14.
55. Langer's affective symbols along with how her work opens up this theory of punctive listening answer what John Durham Peters called for when he remarked: "We need a better name for the infrastructural aesthetics and ethics of being alive with others in the cosmos. Currently *nonverbal* signifies the remainder that is left when you take away language from human communication, but it ignores the meaningfulness found in nonhuman nature" (Peters, *The Marvelous Clouds*, 380).
56. Hoy, interview by author, 18–22.
57. To watch this video, visit "Peacock Spider Dances to YMCA," https://www.youtube.com/watch?v=xYIUFEQeh3g.
58. Hoy, interview by author, 17.
59. Hoy, interview by author, 21.
60. Hoy, interview by author, 18, 11.
61. Hoy, interview by author, 23. To consult Hoy's research, see Bentley and Hoy, "Neurobiology of Cricket Song"; Elias et al., "Courting Male Jumping Spider"; Bruce et al., "Modular Visual System of a Jumping Spider"; Zhou et al., "Outsourced Hearing in an Orb-Weaving Spider." See also Dona et al., "Do Bumble Bees Play?"
62. Hoy, interview by author, 11–12.
63. Prum, *Evolution of Beauty*, 4–5.
64. Prum, *Evolution of Beauty*, 5, 3.
65. Poole, "Orientation."
66. Prum, interview by author, 27.
67. Prum, interview by author, 14.
68. Prum, *Evolution of Beauty*, 14.
69. Prum, interview by author, 36.
70. Prum, interview by author, 8.
71. Brennan, interview by author, 6; emphasis added.
72. This posthumanism is most like philosopher Timothy Morton's in *All Art Is Ecological* and is distinct from others that set out simply to theorize that the human is always already dissolved into the larger whole.
73. Brennan, interview by author, 9.

74. Brennan, interview by author, 10. This work can be read in Brennan, Cowart, and Orbach's "Functional Clitoris in Dolphins" and Orbach et al.'s "Asymmetric and Spiraled Genitalia."
75. Brennan, interview by author, 9.
76. Brennan, interview by author, 24.
77. Brennan, interview by author, 5.
78. Coyne, qtd. in Radiolab, "The Beauty Puzzle."
79. Brennan, interview by author, 21.
80. Brennan, interview by author, 15.
81. Brennan, interview by author, 15.
82. Brennan, interview by author, 15.
83. Hawhee, *Rhetoric in Tooth and Claw*, 4.
84. Prum, interview by author, 32.
85. Prum, interview by author, 24.
86. Eaton et al., "How Gender and Race Stereotypes Impact."
87. Brennan, interview by author, 22.
88. Brennan, interview by author, 22. Later, in an interview with *Current Biology*, Brennan expressed that she wished she would have taken a different approach on this front earlier in her career: "Being a Hispanic female and an immigrant, I ignored microaggressions during most of my academic career. I thought that I should just keep my head down and work twice as hard as everyone else, so that nobody could say that I did not belong. Now I know that I should have been speaking up, not to call anyone out but to invite them into a conversation that could make it easier for the next generations of BIPOC scientists." Brennan, "Q&A," R1065.
89. Prum, interview by author, 23.
90. Prum, interview by author, 23–24.
91. Brennan, interview by author, 16.
92. Hoy, interview by author, 28.
93. Peters, *The Marvelous Clouds*, 382.
94. Kimmerer, *Braiding Sweetgrass*, 39–40.
95. Kimmerer, *Braiding Sweetgrass*, 41.
96. Kimmerer, *Braiding Sweetgrass*, 42.
97. Kimmerer, *Braiding Sweetgrass*, 47.

Conclusion

1. Hoy, interview by author, 27–28.
2. Hoy, interview by author, 43, 29.
3. Hoy, interview by author, 29.
4. Hsu, *Constellating Home*, 11, 140.
5. Peters, *The Marvelous Clouds*, 91.
6. For such studies of listening in rhetorics of health and medicine, see the work of Kristin Marie Bivens.
7. Fournet, interview by author, 20.

Bibliography

Ahmed, Sara. "Happy Objects." In *The Affect Theory Reader*, edited by Melissa Gregg and Gregory J. Seigworth, 29–51. Duke University Press, 2010.
Amundson, Ron. *The Changing Role of the Embryo in Evolutionary Thought: Roots of Evo-Devo*. Cambridge University Press, 2012.
Aristotle. *Rhetoric*. Translated by W. Rhys Roberts. Modern Library, 1954.
Atwill, Janet M. "Introduction: Finding a Home or Making a Path." In *Perspectives on Rhetorical Invention*, edited by Janet M. Atwill and Janice M. Lauer, xi–xxi. University of Tennessee Press, 2003.
Ayto, John. *Dictionary of Word Origins*. Arcade, 1991.
Bacon, Francis. *The New Organon*. 1620. Edited by Lisa Jardine and Michael Silverthorne. Cambridge University Press, 2000.
Barad, Karen. *Meeting the Universe Halfway: Quantum Physics and the Entanglement of Matter and Meaning*. Duke University Press, 2007.
Barnett, Joshua Trey. *Mourning in the Anthropocene: Ecological Grief and Earthly Coexistence*. Michigan State University Press, 2022.
Barthes, Roland. *Camera Lucida: Reflections on Photography*. Translated by Richard Howard. Hill and Wang, 1981.
Barthes, Roland. "The Grain of the Voice." In *Image-Music-Text*. Translated by Stephen Heath, 179–89. Hill & Wang, 1977.
Bar-Yam, Yaneer. "Concepts: Emergence." New England Complex Systems Institute. 2011. https://necsi.edu/emergence.
Bar-Yam, Yaneer. "Concepts: Scale." New England Complex Systems Institute. 2011. https://necsi.edu/scale.
The Beautiful Brain: The Drawings of Santiago Ramón y Cajal. MIT Museum, 2018.
Bedau, Mark A., and Paul Humphreys, eds. *Emergence: Contemporary Readings in Philosophy and Science*. MIT Press, 2008.
Ben-Ari, Elia T. "A Throbbing in the Air." *BioScience* 49, no. 5 (1999): 353–58.
Bentley, David, and Ronald R. Hoy. "The Neurobiology of Cricket Song." *Scientific American* 231, no. 2 (1974): 34–45. https://www.jstor.org/stable/10.2307/24950141.

Berkelhammer, Max. "There Are Signals in the Signals." Presentation at the Rhetoric Society of America Institute, University Park, PA, May 22, 2023.

Berkenkotter, Carol, and Thomas N. Huckin. *Genre Knowledge in Disciplinary Communication: Cognition/Culture/Power*. Routledge, 1994.

Berlant, Lauren. "Cruel Optimism." *differences: A Journal of Feminist Cultural Studies* 17, no. 3 (2006): 20–36. doi:10.1215/10407391-2006-009.

Berlin, James A. "Contemporary Composition: The Major Pedagogical Theories." *College English* 44, no. 8 (1982): 765–77. doi:10.2307/377329.

Berthoff, Ann E. "The Intelligent Eye and the Thinking Hand." In *Landmark Essays on Writing Process*, edited by Sondra Perl, 107–12. Hermagoras Press, 1994.

"Best Books of 2019." *The Washington Post*. November 21, 2019. https://www.washingtonpost.com/graphics/2019/entertainment/books/best-books-of-2019/.

"Best Sellers, Hardcover Fiction." *The New York Times*. July 28, 2019. https://www.nytimes.com/books/best-sellers/2019/07/28/hardcover-fiction/.

Bivens, Kristin Marie. "A Neonatal Intensive Care Unit (NICU) Soundscape: Physiological Monitors, Rhetorical Ventriloquism, and Earwitnessing." *Rhetoric of Health & Medicine* 2, no. 1 (2019): 1–32. doi:10.5744/rhm.2019.1001.

Bivens, Kristin Marie. "Rhetorically Listening for Microwithdrawals of Consent in Research Practice." In *Methodologies for the Rhetoric of Health & Medicine*, edited by Lisa Melonçon and J. Blake Scott, 138–56. Routledge, 2018.

Bjørkdahl, Kristian, and Alex C. Parrish. *Rhetorical Animals: Boundaries of the Human in the Study of Persuasion*. Lexington Books, 2018.

Booher, Amanda K., and Julie Jung, eds. *Feminist Rhetorical Science Studies: Human Bodies, Posthumanist Worlds*. Southern Illinois University Press, 2018.

Borgia, Gerald, and Gregory F. Ball. Review of *The Evolution of Beauty: How Darwin's Forgotten Theory of Mate Choice Shapes the Animal World—and Us*, by Richard O. Prum. *Animal Behaviour* 137 (2018): 187–88. https://doi.org/10.1016/j.anbehav.2017.12.010.

Boyle, Casey. "Aesthetics of Witnessing [Bears] in Late Humanity." *Angelaki* 28, no. 4 (2023): 45–60. https://doi.org/10.1080/0969725X.2023.2233799.

Brennan, Patricia. Interview by Megan Poole. November 22, 2019.

Brennan, Patricia. "Q&A: Patricia Brennan." *Current Biology* 30 (2020): R1063–R1066.

Brennan, Patricia. "Why I Study Duck Genitalia." *Slate*, April 2, 2013. https://slate.com/technology/2013/04/duck-penis-controversy-nsf-is-right-to-fund-basic-research-that-conservatives-misrepresent.html.

Brennan, Patricia L. R., Jonathan R. Cowart, and Dara N. Orbach. "Evidence of a Functional Clitoris in Dolphins." *Current Biology* 32 (2022): R24–R26.

Brennan, Patricia L. R., and Richard O. Prum. "The Limits of Sexual Conflict in the Narrow Sense: New Insights from Waterfowl Biology." *Philosophical Transactions of the Royal Society* 367 (2012): 2324–38. https://doi.org/10.1098/rstb.2011.0284.

Brody, Jane E. "Scientist at Work: Katy Payne; Picking Up Mammals' Deep Notes." *The New York Times*, November 9, 1993. www.nytimes.com/1993/11/09/science/scientist-at-work-katy-payne-picking-up-mammals-deep-notes.html.

brown, adrienne maree. *Emergent Strategy: Shaping Change, Changing Worlds*. AK Press, 2017.

Bruce, Margaret, Daniel Daye, Skye M. Long, Alex M. Winsor, Gil Menda, Ronald R. Hoy, and Elizabeth M. Jakob. "Attention and Distraction in the Modular Visual System of a Jumping Spider." *Journal of Experimental Biology* 224 (2021): 1–5. https://doi.org/10.1242/jeb.231035.

Buehl, Jonathan. *Assembling Arguments: Multimodal Rhetoric and Scientific Discourse*. University of South Carolina Press, 2016.
Burke, Kenneth. *Counter-Statement*. 1931. 3rd ed. University of California Press, 1968.
Burke, Kenneth. *Language as Symbolic Action: Essays on Life, Literature, and Method*. University of California Press, 1966.
Burke, Kenneth. *Permanence and Change: An Anatomy of Purpose*. 1935. 3rd ed. Berkeley: University of California Press, 1984.
Butchart, Garnet C. "The Communicology of Roland Barthes' *Camera Lucida*: Reflections on the Sign-Body Experience of Visual Communication." *Visual Communication* 15, no. 2 (2016): 199–219. https://doi.org/10.1177/1470357215624308.
Campt, Tina M. *Listening to Images*. Duke University Press, 2017.
Ceccarelli, Leah. *Shaping Science with Rhetoric: The Cases of Dobzhansky, Schrödinger, and Wilson*. University of Chicago Press, 2001.
Ceraso, Steph. "(Re)Educating the Senses: Multimodal Listening, Bodily Learning, and the Composition of Sonic Experiences." *College English* 77, no. 2 (2014): 102–23.
Charmaz, Kathy. *Constructing Grounded Theory*. 2006. 2nd ed. Los Angeles: SAGE, 2014.
Chuh, Kandice. *The Difference Aesthetics Makes: On the Humanities "After Man."* Duke University Press, 2019.
Cipolla, Cyd, Kristina Gupta, David A. Rubin, and Angela Willey, eds. *Queer Feminist Science Studies*. University of Washington Press, 2017.
Clary-Lemon, Jennifer. *Nestwork: New Material Rhetorics for Precarious Species*. Pennsylvania State University Press, 2023.
Condit, Celeste Michelle. *The Meanings of the Gene: Public Debates About Human Heredity*. University of Wisconsin Press, 1999.
Cooper, Marilyn M. *The Animal Who Writes: A Posthumanist Composition*. University of Pittsburgh Press, 2019.
Cordey, Huw, dir. *Dancing with the Birds*. Netflix, 2019.
Cornell Lab of Ornithology. "Club-Winged Manakin Dance." YouTube, July 13, 2016. Video. https://www.youtube.com/watch?v=tSHjhCN6NC0.
Cornwall, Warren. "In Hot Water." *Scientific American* 363, no. 6426 (2019): 442–45. https://doi.org/10.1126/science.363.6426.442.
The Countless Aspects of Beauty. Exhibit, May 26, 2018–December 8, 2019. National Archaeological Museum, Athens, Greece.
Coyne, Jerry. "A New Review (and Critique) of Richard Prum's Book on Sexual Selection." *Why Evolution Is True* (blog), August 30, 2018. https://whyevolutionistrue.wordpress.com/2018/08/30/a-new-review-and-critique-of-richard-prums-book-on-sexual-selection/.
Coyne, Jerry. "Which Book Would You Hurl Across the Room?" *Why Evolution Is True* (blog), September 2, 2018. https://whyevolutionistrue.wordpress.com/2018/09/02/which-book-would-your-hurl-across-the-room/.
Damasio, Antonio. *The Feeling of What Happens: Body and Emotion in the Making of Consciousness*. Mariner Books, 2000.
Danto, Arthur C. *The Abuse of Beauty: Aesthetics and the Concept of Art*. Open Court, 2003.
Danto, Arthur C. Foreword to *Mind: An Essay on Human Feeling*, by Susanne K. Langer. Abridged by Gary Van Den Heuvel. Johns Hopkins University Press, 1988.
Daston, Lorraine. "The Naturalistic Fallacy Is Modern." *Isis*, no. 105 (2014): 579–87. https://doi.org/10.1086/678173.

Daston, Lorraine, and Peter Galison. *Objectivity*. Zone Books, 2010.
Davis, Diane. "Some Reflections on the Limit." *Rhetoric Society Quarterly* 47, no. 3 (2017): 275–84.
De Brabandere, Sabine, and Science Buddies. "What Do You Hear Underwater?" *Scientific American*, June 27, 2019. https://www.scientificamerican.com/article/what-do-you-hear-underwater.
De Vaan, Michiel. *Etymological Dictionary of Latin and the Other Italic Languages*. Brill, 2008.
Dona, Hiruni Samadi Galpayage, Cwyn Solvi, Amelia Kowalewska, Kaarle Mäkelä, HaDi MaBouDi, and Lars Chittka. "Do Bumble Bees Play?" *Animal Behaviour* 194 (2022): 239–51. https://doi.org/10.1016/j.anbehav.2022.08.013.
Druschke, Caroline Gottschalk, Nedra Reynolds, Jenna Morton-Aiken, Ingrid E. Lofgren, Nancy E. Karraker, and Scott R. McWilliams. "Better Science Through Rhetoric: A New Model and Pilot Program for Training Graduate Student Science Writers." *Technical Communication Quarterly* 27, no. 2 (2018): 175–90. https://doi.org/10.1080/10572252.2018.1425735.
Dryden, Donald. "Susanne K. Langer." *Dictionary of Literary Biography*. Vol 270, *American Philosophers Before 1950*, edited by Philip B. Dematteis and Leemon B. McHenry, 189–99. Gale Research, 2002.
Eaton, Asia A., Jessica F. Saunders, Ryan K. Jacobson, and Keon West. "How Gender and Race Stereotypes Impact the Advancement of Scholars in STEM: Professors' Biased Evaluations of Physics and Biology Post-Doctoral Candidates." *Sex Roles* 82 (2020): 127–41. https://doi.org/10.1007/s11199-019-01052-w.
Edbauer, Jenny. "Unframing Models of Public Distribution: From Rhetorical Situation to Rhetorical Ecologies." *Rhetoric Society Quarterly* 35, no. 4 (2005): 5–24. https://doi.org/10.1080/02773940509391320.
Eidsheim, Nina Sun. *Sensing Sound: Singing and Listening as Vibrational Practice*. Duke University Press, 2015.
Eisner, Thomas. *For Love of Insects*. Harvard University Press, 2005.
Elephant Listening Project. "About Us." https://www.elephantlisteningproject.org/about-elp/.
Elias, Damian O., Andrew C. Mason, Wayne P. Maddison, and Ronald R. Hoy. "Seismic Signals in a Courting Male Jumping Spider (Araneae: Salticidae)." *Journal of Experimental Biology* 206 (2003): 4029–39. https://doi.org/10.1242/jeb.00634.
Elliott, Wendy, and Mark Simmonds. "Whales in Hot Water? The Impact of a Changing Climate on Whales, Dolphins and Porpoises: A Call for Action." *World Wildlife Federation* (2007). https://de.whales.org/wp-content/uploads/sites/4/2018/07/whales-in-hot-water.pdf.
Emlen, Stephen T., and Peter H. Wrege. "The Role of Kinship in Helping Decisions Among White-Fronted Bee-Eaters." *Behavioral Ecology and Sociobiology* 23 (1988): 305–15. https://doi.org/10.1007/BF00300577.
Emlen, Stephen T., Peter H. Wrege, and Michael S. Webster. "Cuckoldry as a Cost of Polyandry in the Sex-Role-Reversed Wattled Jacana." *Proceedings of the Royal Society* 265, no. 1413 (1998): 2359–64. http://doi.org/10.1098/rspb.1998.0584.
Fahnestock, Jeanne. *Rhetorical Figures in Science*. Oxford University Press, 2002.
Fournet, Michelle. Interview by Megan Poole. September 24, 2021.
Fournet, Michelle E. H., Christine M. Gabriele, David C. Culp, Fred Sharpe, David K. Mellinger, and Holger Klinck. "Some Things Never Change: Multi-Decadal Stability in Humpback Whale Calling Repertoire on Southeast Alaskan Foraging Grounds." *Scientific Reports* 8 (2018): 1–13. https://doi.org/10.1038/s41598-018-31527-x.

Fournet, Michelle E., Leanna P. Matthews, and Christine Gabriele. "The Impact of the 'Anthropause' on the Communication and Acoustic Habitat of Southeast Alaskan Humpback Whales." *Journal of the Acoustical Society of America* 136 (2021). https://doi.org/10.1121/10.0005328.

Fournet, Michelle E., Andy Szabo, and David K. Mellinger. "Repertoire and Classification of Non-Song Calls in Southeast Alaskan Humpback Whales (*Megaptera novaeangliae*). *Journal of the Acoustical Society of America* 137 (2015): 1–10. https://doi.org/10.1121/1.4904504.

Fricker, Miranda. "Intuition and Reason." *Philosophical Quarterly* 45, no. 179 (1995): 181–89. https://www.jstor.org/stable/2220414.

Fricker, Miranda. "Why '*Female* Intuition'?" *Women's Philosophy Review*, no. 15 (1996): 36–44.

Frost, Robert. "The Tuft of Flowers." 1915. Poetry Foundation. https://www.poetryfoundation.org/poems/44275/the-tuft-of-flowers.

Gagne, Emma, Betzi Perez-Ortega, Andrew P. Hendry, Gabriel Melo-Santos, Sam F. Walmsley, Manali Rege-Colt, Maia Austin, and Laura J. May-Collado. "Dolphin Communication During Widespread Systematic Noise Reduction—A Natural Experiment amid COVID-19 Lockdowns." *Frontiers in Remote Sensing* 3 (2022): 1–14. https://doi.org/10.3389/frsen.2022.934608.

Gardner, Howard. *Frames of Mind: The Theory of Multiple Intelligences*. 1983. 20th Anniversary ed. New York: Basic Books, 2004.

Garland, Ellen C., Anne W. Goldizen, Melinda L. Rekdahl, Rochelle Constantine, Claire Garrigue, Nan Daeschler Hauser, M. Michael Poole, Jooke Robbins, and Michael J. Noad. "Dynamic Horizontal Cultural Transmission of Humpback Whale Song at the Ocean Basin Scale." *Current Biology* 21 (2011): 687–91. https://doi.org/10.1016/j.cub.2011.03.019.

Gautier, Ana María Ochoa. *Aurality: Listening and Knowledge in Nineteenth-Century Columbia*. Duke University Press, 2014.

Gigante, Maria E. *Introducing Science Through Images: Cases of Visual Popularization*. University of South Carolina Press, 2018.

Global Conservation. "African Forest Elephant." Last modified 2023. https://globalconservation.org/endangered-species/african-forest-elephant.

Gomes-Selman, Jonathan M., Nikita Demir, Peter Wrege, and Andreas Paepcke. "Curriculum Learning to Handle Extreme Class Imbalance for Acoustic Modeling of Forest Elephant Calls." *IEEE International Conference on Machine Learning and Applications* (2021): 1022–28. https://doi.org/10.1109/ICMLA52953.2021.00168.

Goodale, Greg. *The Rhetorical Invention of Man: A History of Distinguishing Humans from Other Animals*. Lexington Books, 2015.

Graham, S. Scott. "The Promise and Peril of Scientific Science Studies." *Theory & Event* 21, no. 2 (2018): 529–38. https://doi.org/10.1353/tae.2018.0027.

Graham, S. Scott. *Where's the Rhetoric? Imagining a Unified Field*. Ohio State University Press, 2020.

Gross, Alan. *The Rhetoric of Science*. Harvard University Press, 1996.

Gross, Alan. *The Scientific Sublime: Popular Science Unravels the Mysteries of the Universe*. Oxford University Press, 2018.

Gross, Alan G., and Joseph E. Harmon. *Science from Sight to Insight: How Scientists Illustrate Meaning*. University of Chicago Press, 2013.

Gross, Daniel M. *Being-Moved: Rhetoric as the Art of Listening*. University of California Press, 2020.

Grosz, Elizabeth. *The Incorporeal: Ontology, Ethics, and the Limits of Materialism*. Columbia University Press, 2018.

Gugleilmi, Giorgia. "The Biologist Using Insect Eggs to Overturn Evolutionary Doctrine." *Nature*, July 3, 2019. https://www.nature.com/articles/d41586-019-02040-6.

Guinee, Linda N., and Katharine B. Payne. "Rhyme-like Repetitions in Songs of Humpback Whales." *Ethology* 79, no. 4 (1988): 295–308. https://doi.org/10.1111/j.1439-0310.1988.tb00718.x.

Gumbs, Alexis Pauline. *Undrowned: Black Feminist Lessons from Marine Mammals*. AK Press, 2020.

Haraway, Donna J. *Modest_Witness@Second_Millennium.FemaleMan©_Meets_OncoMouse™: Feminism and Technoscience*. Routledge, 1997.

Haraway, Donna. "Situated Knowledges: The Science Question in Feminism and the Privilege of Partial Perspective." *Feminist Studies* 14, no. 3 (1988): 575–99. https://doi.org/10.2307/3178066.

Haraway, Donna J. *Staying with the Trouble: Making Kin in the Chthulucene*. Duke University Press, 2016. Harding, Sandra. *Objectivity and Diversity: Another Logic of Scientific Research*. University of Chicago Press, 2015.

Harding, Sandra. *Whose Science? Whose Knowledge? Thinking from Women's Lives*. Cornell University Press, 1991.

Harrison, Michaela. *Whale Whispering* (blog). https://www.michaelaharrison.org/whale-whispering-1.

Hawhee, Debra. *Moving Bodies: Kenneth Burke at the Edges of Language*. University of South Carolina Press, 2012.

Hawhee, Debra. *Rhetoric in Tooth and Claw: Animals, Language, Sensation*. University of Chicago Press, 2017.

Hawhee, Debra. "Rhetoric's Sensorium." *Quarterly Journal of Speech* 101, no. 1 (2015): 2–17. https://doi.org/10.1080/00335630.2015.995925.

Hawhee, Debra. *A Sense of Urgency: How the Climate Crisis Is Changing Rhetoric*. University of Chicago Press, 2023.

Hawk, Byron. "Sound: Resonance as Rhetorical." *Rhetoric Society Quarterly* 48, no. 3 (2018): 315–23. https://doi.org/10.1080/02773945.2018.1454219.

Hedwig, Daniela. Interview by Megan Poole. July 29, 2021.

Heller, Marielle, dir. *A Beautiful Day in the Neighborhood*. TriStar Pictures, 2019.

Helms, Jason. *Rhizcomics: Rhetoric, Technology, and New Media Composition*. University of Michigan Press, 2017. https://doi.org/10.3998/mpub.7626373.

Herman, Louis M. "The Multiple Functions of Male Song Within the Humpback Whale (*Megaptera novaeangliae*) Mating System: Review, Evaluation, and Synthesis." *Biological Reviews* 92 (2017): 1795–818. https://doi.org/10.1111/brv.12309.

Hoffman, Julian. "A Language of Listening." In *Kinship: Partners*. Vol. 3, edited by Gavin Van Horn, Robin Wall Kimmerer, and John Hausdoerffer, 52–61. Center for Humans and Nature Press, 2021.

Hoffmeyer, Jesper. *Biosemiotics: An Examination into the Signs of Life and the Life of Signs*. University of Chicago Press, 2008.

Horowitz, Alexandra. *On Looking: A Walker's Guide to the Art of Observation*. Simon & Schuster, 2014.

Hoy, Ronald R. Interview by Megan Poole. July 30, 2021.

Hsu, V. Jo. *Constellating Home: Trans and Queer Asian American Rhetorics.* Ohio State University Press, 2022.

Jack, Jordynn. *Autism and Gender: From Refrigerator Mothers to Computer Geeks.* University of Illinois Press, 2014.

Johnson, Mark. *The Aesthetics of Meaning and Thought: The Bodily Roots of Philosophy, Science, Morality, and Art.* University of Chicago Press, 2018.

Jones, Natasha. "Coalitional Learning in the Contact Zones: Inclusion and Narrative Inquiry in Technical Communication and Composition Studies." *College English* 82, no. 5 (2020): 515–26.

Jones, Natasha, and Rebecca Walton. "Narratives to Foster Critical Thinking About Diversity and Social Justice." In *Key Theoretical Frameworks: Teaching Technical Communication in the Twenty-First Century,* edited by Angela Haas and Michelle F. Eble, 241–67. Utah State University Press, 2018.

Kassamali-Fox, Ayshah, Fredrik Christiansen, Laura J. May-Collado, Eric A. Ramos, and Beth A. Kaplin. "Tour Boats Affect the Activity Patterns of Bottlenose Dolphins (*Tursiops truncates*) in Bocas del Toro, Panama." *PeerJ* (2020): 1–22. https://doi.org/10.7717/peerj.8804.

Keeling, Diane M. "Feral Rhetoric: Common Sense Animals and Metaphorical Beasts." *Rhetoric Society Quarterly* 47, no. 3 (2017): 229–37. https://doi.org/10.1080/02773945.2017.1309905.

Keller, Evelyn Fox. *A Feeling for the Organism: The Life and Work of Barbara McClintock.* Holt, 1983.

Keller, Evelyn Fox. *Reflections on Gender and Science.* 10th Anniversary ed. Yale University Press, 1996.

Kemp, Martin. *Structural Intuitions: Seeing Shapes in Art and Science.* University of Virginia Press, 2016.

Kennedy, George A. "A Hoot in the Dark: The Evolution of General Rhetoric." *Philosophy & Rhetoric* 25, no. 1 (1992): 1–21. http://www.jstor.org/stable/40238276.

Kepes, György. *Art Education for Scientist and Engineer.* CAVS Special Collections. Institute Archives & Special Collections, MIT Libraries.

Kepes, György. Letter to Susanne K. Langer, December 17, 1958. Susanne K. Langer Papers. Houghton Library, Harvard University.

Kepes, György. Letter to Susanne K. Langer, January 21, 1959. Susanne K. Langer Papers. Houghton Library, Harvard University.

Kepes, György. Letter to Susanne K. Langer. August 28, 1959. Susanne K. Langer Papers. Houghton Library, Harvard University.

Kepes, György. Letter to Susanne K. Langer. January 28, 1962. Susanne K. Langer Papers. Houghton Library, Harvard University.

Kimmerer, Robin Wall. *Braiding Sweetgrass: Indigenous Wisdom, Scientific Knowledge, and the Teaching of Plants.* Milkweed Editions, 2013.

Kimmerer, Robin Wall. "Epilogue—Attention, Curiosity, Play, Gratitude: Practices of Kinship." In *Kinship: Persons,* edited by Gavin Van Horn, Robin Wall Kimmerer, and John Hausdoerffer, 4:127–47. Center for Humans and Nature Press, 2021.

Knuuttila, Sirkka. "*L'effet de réel* Revisited: Barthes and the Affective Image." *Sign Systems Studies* 36, no. 1 (2008): 113–36.

Kohn, Eduardo. *How Forests Think: Toward an Anthropology Beyond the Human.* University of California Press, 2013.

Kristeva, Julia. *Revolution in Poetic Language.* Translated by Margaret Waller. Columbia University Press, 1984.
Kuhn, Thomas S. "Comments on the Relations of Science and Art." In *The Essential Tension: Selected Studies in Scientific Tradition and Change,* 340–51. University of Chicago Press, 1977.
Kuhn, Thomas S. *The Structure of Scientific Revolutions.* 4th ed. University of Chicago Press, 2012.
Langer, Susanne K. "Artistic Perception & Natural Light." Susanne K. Langer Papers. Houghton Library, Harvard University.
Langer, Susanne K. *Feeling and Form: A Theory of Art.* Scribner, 1953.
Langer, Susanne K. "The Great Shift: Instinct to Intuition." Susanne K. Langer Papers. Houghton Library, Harvard University.
Langer, Susanne K. Letter to György Kepes. December 31, 1958. Susanne K. Langer Papers. Houghton Library, Harvard University.
Langer, Susanne K. Letter to György Kepes. September 1, 1959. Susanne K. Langer Papers. Houghton Library, Harvard University.
Langer, Susanne K. Letter to György Kepes. February 6, 1962. Susanne K. Langer Papers. Houghton Library, Harvard University.
Langer, Susanne K. Letter to György Kepes. February 27, 1974. Susanne K. Langer Papers. Houghton Library, Harvard University, AM 3110, box 1, folder 3.
Langer, Susanne K. Letter to Jerome Archer. June 22, 1956. Susanne K. Langer Papers. Houghton Library, Harvard University.
Langer, Susanne K. *Mind: An Essay on Human Feeling.* 3 vols. Johns Hopkins Press, 1967–82.
Langer, Susanne K. *Mind: An Essay on Human Feeling.* Abridged edition. Edited by Gary Van Den Heuvel. Johns Hopkins University Press, 1988.
Langer, Susanne K. "The Mind in Its Own Image." Susanne K. Langer Papers. Houghton Library, Harvard University.
Langer, Susanne K. *Philosophy in a New Key: A Study in the Symbolism of Reason, Rite, and Art.* 3rd ed. Harvard University Press, 1957.
Larson, Elaine. "The History of the Ivory Trade." *National Geographic.* https://education.nationalgeographic.org/resource/history-ivory-trade/.
Lee, A. Robert. "Survivance Memories: The Poetry of Carter Revard." In *Survivance: Narratives of Native Presence,* edited by Gerald Vizenor, 333–54. University of Nebraska Press, 2008.
Levine, George. "Why Beauty Matters." *Victorian Literature and Culture* 47, no. 1 (2018): 243–65. https://doi.org/10.1017/S106015031800147X.
Lewandowski, Annie. Interview by Megan Poole. July 28, 2021.
Lewandowski, Annie. Interview by Theresa Wong. February 2021.
Lewandowski, Annie, Kyle McDonald, and Amy Rubin. *Siren—Listening to Another Species on Earth.* Exhibit, September 23–25, 2021. Herbert F. Johnson Museum of Art, Ithaca, NY.
Leys, Ruth. *The Ascent of Affect: Genealogy and Critique.* University of Chicago Press, 2017.
Liboiron, Max. *Pollution Is Colonialism.* Duke University Press, 2021.
Limón, Ada. "The Whale and the Waltz Inside of It." In *Bright Dead Things.* Milkweed Editions, 2015.
Lindsey, Rebecca, and LuAnn Dahlman. "Climate Change: Global Temperature." Climate.gov. January 18, 2023. https://www.climate.gov/news-features/understanding-climate/climate-change-global-temperature.

Lopez, Jaime. "Night Chirps in Costa Rica: Singing Mice and Clicking Geckos." *Costa Rica Star*, n.d. https://news.co.cr/night-chirps-costa-rica-singing-mice-clicking-geckos/12580/#google_vignette.

Lorenz, Konrad. *King Solomon's Ring: New Light on Animal Ways*. Thomas Y. Crowell Company, 1952. Martino Fine Books, 2020.

Lyon, Arabella. "Susanne K. Langer: Mother and Midwife at the Rebirth of Rhetoric." In *Reclaiming Rhetorica: Women in the Rhetorical Tradition*, edited by Andrea A. Lunsford, 265–84. University of Pittsburgh Press, 1995.

Mandoki, Katya. *The Indispensable Excess of the Aesthetic: Evolution of Sensibility in Nature*. Lexington Books, 2015.

Massumi, Brian. *Parables for the Virtual: Movement, Affect, Sensation*. Duke University Press, 2002.

Massumi, Brian. *Semblance and Event: Activist Philosophy and the Occurrent Arts*. MIT Press, 2013.

May-Collado, Laura J. "Changes in Whistle Structure of Two Dolphin Species During Interspecific Associations." *Ethology* 116 (2010): 1065–74. https://doi.org/10.1111/j.1439-0310.2010.01828.x.

May-Collado, Laura J. Interview by Megan Poole. August 2, 2021.

May-Collado, Laura J., Ingi Agnarsson, and Douglas Wartzok. "Phylogenetic Review of Tonal Sound Production in Whales in Relation to Sociality." *BMC Evolutionary Biology* 7, no. 136 (2007): 1–20. https://doi.org/10.1186/1471-2148-7-136.

May-Collado, Laura J., Ingi Agnarsson, and Douglas Wartzok. "Reexamining the Relationship Between Body Size and Tonal Signals Frequency in Whales: A Comparative Approach Using a Novel Phylogeny." *Marine Mammal Science* 23, no. 3 (2007): 524–52. https://doi.org/10.1111/j.1748-7692.2007.02250.x.

May-Collado, Laura J., and Douglas Wartzok. "A Comparison of Bottlenose Dolphin Whistles in the Atlantic Ocean: Factors Promoting Whistle Variation." *Journal of Mammalogy* 89, no. 5 (2008): 1229–40.

McGreavy, Bridie, Justine Wells, George F. McHendry Jr., and Samantha Senda-Cook. *Tracing Rhetoric and Material Life: Ecological Approaches*. Palgrave, 2018.

McKittrick, Katherine. *Dear Science and Other Stories*. Duke University Press, 2021.

McLeish, Tom. *The Poetry and Music of Science: Comparing Creativity in Science and Art*. Oxford University Press, 2019.

McQuay, Bill. Interview by Megan Poole. September 24, 2021.

McQuay, Bill. "Night at Dzanga Bai Clearing." *NPR*, 2015. https://soundcloud.com/npr/night-at-dzanga-bai-clearing.

McQuay, Bill, and Christopher Joyce. "It Took a Musician's Ear to Decode the Complex Song in Whale Calls." In *Close Listening: Decoding Nature Through Sound from NPR*, August 6, 2015. https://www.npr.org/2015/08/06/427851306/it-took-a-musicians-ear-to-decode-the-complex-song-in-whale-calls.

Melville, Herman. *Moby-Dick*. 1851. Edited by John Bryant and Haskell Springer. London: Pearson, 2007.

Miller, Carolyn. "The Appeal(s) of Latour." *Rhetoric Society Quarterly* 47, no. 5 (2017): 454–62.

Moore, Michael J. *We Are All Whalers: The Plight of Whales and Our Responsibility*. University of Chicago Press, 2021.

Morton, Timothy. *All Art Is Ecological*. Penguin, 2020.

Muckelbauer, John. "Asignification." In *A New Handbook of Rhetoric: Inverting the Classical Vocabulary*, edited by Michele Kennerly. Pennsylvania State University Press, 2021.

Muckelbauer, John. *The Future of Invention: Rhetoric, Postmodernism, and the Problem of Change*. State University of New York Press, 2008.

Nancy, Jean-Luc. *Listening*. Translated by Charlotte Mandell. Fordham University Press, 2007.

National Aeronautics and Space Administration (NASA). "How Long Have Sea Levels Been Rising?" n.d. Accessed July 24, 2023. https://sealevel.nasa.gov/faq/13/how-long-have-sea-levels-been-rising-how-does-recent-sea-level-rise-compare-to-that-over-the-previous/.

National Oceanic and Atmospheric Administration (NOAA). "U.S. Coastline to See Up to a Foot of Sea Level Rise by 2050." February 15, 2022. https://www.noaa.gov/news-release/us-coastline-to-see-up-to-foot-of-sea-level-rise-by-2050.

National Oceanic and Atmospheric Administration Fisheries. "Whales and Climate Change: Big Risks to the Ocean's Biggest Species." Last modified June 23, 2022. https://www.fisheries.noaa.gov/national/climate/whales-and-climate-change-big-risks-oceans-biggest-species.

Nezhukumatathil, Aimee. *World of Wonders: In Praise of Fireflies, Whale Sharks, and Other Astonishments*. Milkweed Editions, 2020.

Nocek, Adam. "A Feeling for Biological Concepts: Countering the 'Idols of the Laboratory' in the Thought of Susanne K. Langer." In *The Future Whispers: The Bloomsbury Handbook of Susanne K. Langer*, edited by Lona Gaikis, 249–74. Bloomsbury, 2024.

Nutt, David. "Collaboration Showcases Creativity of Whale Songs." *Cornell Chronicle*, June 18, 2019. news.cornell.edu/stories/2019/06/collaboration-showcases-creativity-whale-songs.

Okun, Tema. "Sense of Urgency." (Divorcing) White Supremacy Culture. Last modified June 2023. https://www.whitesupremacyculture.info/urgency.html.

Oliver, Kelly. "Witnessing, Recognition, and Response Ethics." *Philosophy & Rhetoric* 48, no. 4 (2015): 473–93. https://muse.jhu.edu/article/602490.

Oliveros, Pauline. *Deep Listening: A Composer's Sound Practice*. iUniverse, 2005.

Orbach, Dara N., Patricia L. R. Brennan, Brandon P. Hedrick, William Keener, Marc A. Webber, and Sarah L. Mesnick. "Asymmetric and Spiraled Genitalia Coevolve with Unique Lateralized Mating Behavior." *Scientific Reports* 10, no. 3257 (2020): 1–8. https://doi.org/10.1038/s41598-020-60287-w.

Ott, Brian L. "Affect in Critical Studies." In *Oxford Research Encyclopedia of Communication*, July 27, 2017. https://doi.org/10.1093/acrefore/9780190228613.013.56.

Parr, Patrick. "Barbara McClintock (1902–1992): Fighting the Male Establishment." TheHumanist.com, March 21, 2016. https://thehumanist.com/features/articles/barbara-mcclintock-1902-1992-fighting-male-establishment/.

Parrish, Alex C. *The Sensory Modes of Animal Rhetorics: A Hoot in the Light*. New York: Palgrave Macmillan, 2021.

Partridge, Eric. *Origins: A Short Etymological Dictionary of Modern English*. Routledge, 1977.

Patricelli, Gail, Eileen A. Hebets, and Tamra C. Mendelson. Review of *The Evolution of Beauty: How Darwin's Forgotten Theory of Mate Choice Shapes the Animal World—and Us*, by Richard O. Prum. *Evolution* 73, no. 1 (2018): 115–24. https://doi.org/10.1111/evo.13629.

Payne, Katharine B., William R. Langbauer Jr., and Elizabeth M. Thomas. "Infrasonic Calls of the Asian Elephant (*Elephas maximus*)." *Behavioral Ecology and Sociobiology* 18, no. 4 (1986): 297–301. https://www.jstor.org/stable/4599893.

Payne, Katy. Interview by Megan Poole. January 15, 2018.
Payne, Katy. "In the Presence of Elephants and Whales." Interview by Krista Tippett. *The On Being Project*. February 1, 2007. https://onbeing.org/programs/katy-payne-in-the-presence-of-elephants-and-whales/#transcript.
Payne, Katy. "The Progressively Changing Songs of Humpback Whales: A Window on the Creative Process in a Wild Animal." In *The Origins of Music*, edited by Nils L. Wallin, Björn Merker, and Steven Brown, 135–50. MIT Press, 2000.
Payne, Katy. *Silent Thunder: In the Presence of Elephants*. Simon & Schuster, 1998.
Payne, Katy. "Sources of Social Complexity in the Three Elephant Species." In *Animal Social Complexity*, edited by Frans B. M. de Waal and Peter L. Tyack, 57–85. Harvard University Press, 2003.
Payne, Roger. *Among Whales*. Scribner, 1995.
Payne, Roger. *Songs of the Humpback Whale*. CRM Records, 1970.
Payne, Roger, and Scott McVay. "Songs of Humpback Whales." *Science* 173, no. 3997 (1971): 585–97.
Peirce, Charles Sanders. *Peirce on Signs*, edited by James Hoopes. University of North Carolina Press, 1991.
Pender, Kelly. *Techne, from Neoclassicism to Postmodernism: Understanding Writing as a Useful, Teachable Art*. Parlor Press, 2011.
Perez-Ortega, Betzi, Rebecca Daw, Brennan Paradee, Emma Gimbrere, and Laura J. May-Collado. "Dolphin-Watching Boats Affect Whistle Frequency Modulation in Bottlenose Dolphins." *Frontiers in Marine Science* 8 (2021): 1–12. https://doi.org/10.3389/fmars.2021.618420.
Peters, John Durham. *The Marvelous Clouds: Toward a Philosophy of Elemental Media*. University of Chicago Press, 2015.
Peters, John Durham. "Witnessing." *Media, Culture, & Society* 23 (2001): 707–23.
Pink, Sarah. *Doing Sensory Ethnography*. 2nd ed. SAGE, 2015.
Pippin, Robert B. *After the Beautiful: Hegel and the Philosophy of Pictorial Modernism*. University of Chicago Press, 2013.
Plec, Emily, ed. *Perspectives on Human-Animal Communication*. Routledge, 2013.
Polanyi, Michael. *The Tacit Dimension*. 1966. University of Chicago Press, 2009.
Poole, Joyce H., Katharine Payne, William R. Langbauer Jr., and Cynthia J. Moss. "The Social Contexts of Some Very Low Frequency Calls of African Elephants." *Behavioral Ecology and Sociobiology* 22, no. 6 (1988): 385–92. https://jstor.org/stable/4600167.
Poole, Megan. "Orientation: Seeing and Sensing Rhetorically." *Western Journal of Communication* 84, no. 5 (2020): 604–22. https://doi.org/10.1080/10570314.2020.1760341.
Prelli, Lawrence J. *A Rhetoric of Science: Inventing Scientific Discourse*. University of South Carolina Press, 1989.
Prescod-Weinstein, Chanda. "Making Black Women Scientists Under White Empiricism: The Racialization of Epistemology in Physics." *Signs* 45, no. 2 (2020): 421–47. https://doi.org/10.1086/704991.
Propen, Amy. *Visualizing Posthuman Conservation in the Anthropocene*. Ohio State University Press, 2018.
Prum, Richard O. "Anatomy, Physics, and Evolution of Structural Color." In *Bird Coloration: Mechanisms and Measurements*, edited by Geoffrey E. Hill and Kevin J. McGraw, 295–353. Harvard University Press, 2006.

Prum, Richard O. "Duck Sex and the Patriarchy." *The New Yorker*, May 17, 2017. https://www.newyorker.com/tech/annals-of-technology/duck-sex-and-the-patriarchy.

Prum, Richard O. *The Evolution of Beauty: How Darwin's Forgotten Theory of Mate Choice Shapes the Animal World—and Us*. Doubleday, 2017.

Prum, Richard O. Interview by Megan Poole. April 18, 2019.

Prum, Richard O. "The Lande-Kirkpatrick Mechanism Is the Null Model of Evolution by Intersexual Selection: Implications for Meaning, Honesty, and Design in Intersexual Signals." *Perspective* 64, no. 11 (2010): 3085–100. https://doi.org/10.1111/j.1558-5646.2010.01054.x.

Prum, Richard O., and Alan H. Brush. "Which Came First, the Feather or the Bird?" *Scientific American* (2003): 84–93.

Rai, Candice, and Caroline Gottschalk Druschke. *Field Rhetoric: Ethnography, Ecology, and Engagement in the Places of Persuasion*. University of Alabama Press, 2018.

Radiolab. "The Beauty Puzzle." Reported by Robert Krulwich and Bethel Habte, produced by Bethel Habte. Aired February 8, 2019, WYNC Studios. https://www.wnycstudios.org/podcasts/radiolab/articles/beauty-puzzle.

Ratcliffe, Krista. *Rhetorical Listening: Identification, Gender, Whiteness*. Southern Illinois University Press, 2005.

Revard, Carter. *How the Songs Come Down*. Salt, 2005.

Richerme, Lauren Kapalka. "A Deleuzian Reimagining of Susanne Langer's Philosophy: Becoming-Feeling in Music Education." *Music Education Research* 20, no. 3 (2018): 330–41. https://doi.org/10.1080/14613808.2017.1409201.

Rickert, Thomas. *Ambient Rhetoric: The Attunements of Rhetorical Being*. University of Pittsburgh Press, 2013.

Rickert, Thomas. "Toward the Chora: Kristeva, Derrida, and Ulmer on Emplaced Invention." *Philosophy & Rhetoric* 40, no. 3 (2007): 251–73. https://www.jstor.org/stable/25655276.

Rivers, Nathaniel A. "Better Footprints." In *Tracing Rhetoric and Material Life: Ecological Approaches*, edited by Bridie McGreavy, Justine Wells, George F. McHendry Jr., and Samantha Senda-Cook, 169–96. Palgrave Macmillan, 2017.

Robinson, Dylan. *Hungry Listening: Resonant Theory for Indigenous Sound Studies*. University of Minnesota Press, 2020.

Rothenberg, David. *Thousand Mile Song: Whale Music in a Sea of Sound*. Basic Books, 2008.

Rowland, Liz. Interview by Megan Poole. July 30, 2021.

Runia, Eelco. *Moved by the Past: Discontinuity and Historical Mutation*. Columbia University Press, 2014.

Runia, Eelco. "Presence." *History and Theory* 45 (2006): 1–29.

Safina, Carl. *Becoming Wild: How Animal Cultures Raise Families, Create Beauty, and Achieve Peace*. Picador, 2020.

Safina, Carl. *Beyond Words: What Animals Think and Feel*. Picador, 2015.

Sakakibara, Chie. *Whale Snow: Iñupiat, Climate Change, and Multispecies Resilience in Arctic Alaska*. University of Arizona Press, 2020.

Seeley, Thomas D. *Honeybee Democracy*. Princeton University Press, 2010.

Schafer, R. Murray. *The Soundscape: Our Sonic Environment and the Tuning of the World*. 1977. Destiny Books, 1993.

Schulkin, Jay. *Reflections on the Musical Mind: An Evolutionary Perspective*. Princeton University Press, 2013.

Schultz, William. *Cassirer and Langer on Myth: An Introduction*. Garland, 2000.

Scott, John Blake. *Risky Rhetoric: AIDS and the Cultural Practices of HIV Testing*. Southern Illinois University Press, 2003.
Sedgwick, Eve Kosofsky. *Touching Feeling: Affect, Pedagogy, Performativity*. Duke University Press, 2003.
Seegert, Natasha. "Play of Sniffication: Coyotes Sing in the Margins." *Philosophy & Rhetoric* 47, no. 2 (2014): 158–78. https://muse.jhu.edu/article/545810.
Shapin, Steven, and Simon Schaffer. *Leviathan and the Air-Pump: Hobbes, Boyle, and the Experimental Life*. 1985. Princeton University Press, 2011.
Sharpe, Christina. *Ordinary Notes*. Alfred A. Knopf, 2023.
Simonson, Peter. "Reinventing Invention, Again." *Rhetoric Society Quarterly* 44, no. 4 (2014): 299–322. https://doi.org/10.1080/02773945.2014.938862.
Sitar, Ashley, Laura J. May-Collado, Andrew Wright, Erin Peters-Burton, Larry Rockwood, and E. C. M. Parsons. "Tourists' Perspectives on Dolphin Watching in Bocas del Toro, Panama." *Tourism in Marine Environments* 12, no. 2 (2017): 79–94. https://doi.org/10.3727/1544273 16X14820977775343.
Smolen, Ann. "Dance." In *Textbook of Applied Psychoanalysis*, edited by Salman Akhtar and Stuart Twemlow, 379–96. Routledge, 2018.
Stengers, Isabelle. *Thinking with Whitehead: A Free and Wild Creation of Concepts*. Translated by Michael Chase. Harvard University Press, 2011.
Stewart, Kathleen. "Atmospheric Attunements." *Society and Space* 29 (2011): 445–53. https://doi.org/10.1068/d9109.
Stormer, Nathan, and Bridie McGreavy. "Thinking Ecologically About Rhetoric's Ontology: Capacity, Vulnerability, and Resilience." *Philosophy and Rhetoric* 50, no. 1 (2017): 1–25.
Swider, Colin R., Christopher F. Gemelli, Peter H. Wrege, and Susan E. Parks. "Passive Acoustic Monitoring Reveals Behavioural Response of African Forest Elephants to Gunfire Events." *African Journal of Ecology* 60 (2022): 882–94. https://doi.org/10.1111/aje.13070.
TallBear, Kim. "Standing with and Speaking as Faith: A Feminist-Indigenous Approach to Inquiry." *Journal of Research Practice* 10, no. 2 (2014): 1–7.
Taylor, Christie, and Ira Flatow. "How the Humpback Says Hello." *Science Friday*, July 23, 2021. https://www.sciencefriday.com/segments/humpback-documentary-fathom/.
Taylor, Hollis. *Is Birdsong Music? Outback Encounters with an Australian Songbird*. Indiana University Press, 2017.
Thompson, Andrea. "July 2023 Is Hottest Month Ever Recorded on Earth." *Scientific American*, July 27, 2023. https://www.scientificamerican.com/article/july-2023-is-hottest-month-ever-recorded-on-earth/.
Tuhiwai Smith, Linda. *Decolonizing Methodologies: Research and Indigenous Peoples*. 1999. 2nd ed. Zed Books, 2012.
Tuin, Iris van der. "Bergson Before Bergsonism: Traversing 'Bergson's Failing' in Susanne K. Langer's Philosophy of Art." *Journal of French and Francophone Philosophy* 24, no. 2 (2016): 176–202. https://doi.org/10.5195/jffp.2016.776.
Tuin, Iris van der. "Diffraction as a Methodology for Feminist Onto-Epistemology: On Encountering Chantal Chawaf and Posthuman Interpellation." *Parallax* 20, no. 3 (2014): 231–44. https://doi.org/10.1080/13534645.2014.927631.
Uexküll, Jakob von. *A Foray into the Worlds of Animals and Humans*. Translated by Joseph D. O'Neil. University of Minnesota Press, 2010.

Van Dooren, Thom, and Matthew Chrulew, eds. *Kin: Thinking with Deborah Rose Bird*. Duke University Press, 2022.

Van Horn, Gavin. "Kinning: Introducing the Kinning Series." In *Kinship: Planet*. Vol. 1, edited by Gavin Van Horn, Robin Wall Kimmerer, and John Hausdoerffer, 1–11. Center for Humans and Nature Press, 2021.

Vivian, Bradford. *Commonplace Witnessing: Rhetorical Invention, Historical Remembrance, and Public Culture*. Oxford University Press, 2017.

Vuong, Ocean. *On Earth We're Briefly Gorgeous*. Penguin, 2019.

Waldrop, Mitch. "Inside Einstein's Love Affair with 'Lina'—His Cherished Violin." *National Geographic*, February 3, 2017. https://www.nationalgeographic.com/adventure/article/einstein-genius-violin-music-physics-science.

Weilgart, Linda, Hal Whitehead, and Katharine Payne. "A Colossal Convergence." *American Scientist* 84, no. 3 (1996): 278–87. http://www.jstor.org/stable/29775673.

Westling, Louise. *The Logos of the Living World: Merleau-Ponty, Animals, and Language*. Fordham University Press, 2014.

Whitehead, Alfred North. *Science and the Modern World*. 1925. Macmillan, 1967.

Whitehead, Alfred North. *Symbolism: Its Meaning and Effect*. 1927. Reprint, Fordham University Press, 1955.

Whitehead, Hal. "Society and Culture in the Deep and Open Ocean: The Sperm Whale and Other Cetaceans." In *Animal Social Complexity: Intelligence, Culture, and Individualized Societies*, edited by Frans B. M. de Waal and Peter L. Tyack, 444–69. Harvard University Press, 2003.

Whitehead, Hal, and Luke Rendell. *The Cultural Lives of Whales and Dolphins*. University of Chicago Press, 2015.

Williams, Brooke. "Four Turtles." In *Kinship: Persons*. Vol. 4, edited by Gavin Van Horn, Robin Wall Kimmerer, and John Hausdoerffer, 63–69. Center for Humans and Nature Press, 2021.

Won't You Be My Neighbor? Created by Morgan Neville, Trimolo Productions, 2018.

World Wildlife Fund. "The Status of African Elephants." *World Wildlife Magazine*. Winter 2018. https://www.worldwildlife.org/magazine/issues/winter-2018/articles/the-status-of-african-elephants.

Wrege, Peter. Interview by Megan Poole. August 4, 2021.

Wrege, Peter H., Elizabeth D. Rowland, Sara Keen, and Yu Shiu. "Acoustic Monitoring for Conservation in Tropical Forests: Examples from Forest Elephants." *Methods in Ecology and Evolution* 8 (2017): 1292–301. https://doi.org/10.1111/2041-210X.12730.

Wrege, Peter H., Elizabeth D. Rowland, Bruce G. Thompson, and Nikolas Batruch. "Use of Acoustic Tools to Reveal Otherwise Cryptic Responses of Forest Elephants to Oil Exploration." *Conservation Biology* 24, no. 6 (2010): 1578–85. https://doi.org/10.1111/j.1523-1739.2010.01559.x.

Xanthopoulos, Drew, dir. *Fathom*. Back Allie Productions, 2021.

Yong, Ed. *An Immense World: How Animal Senses Reveal the Hidden Realms Around Us*. Random House, 2022.

Zhou, Jian, Junpeg Lai, Gil Menda, Jay A. Stafstrom, Carol I. Miles, Ronald R. Hoy, and Ronald N. Miles. "Outsourced Hearing in an Orb-Weaving Spider That Uses Its Web as an Auditory Sensor." *PNAS* 119, no. 14 (2022): 1–7. https://doi.org/10.1073/pnas.2122789119.

Index

Page numbers followed by an "f" refer to figures.

abstraction, 25, 30, 154n41
aesthesis, 11–13, 25, 53, 65, 68, 79, 90, 110, 123, 135, 150nn36–37; as defined, 7; and sound, 26
aesthetic evolution, 110–11, 112–13, 126, 133; adaptationists, rift between, 123; art for art's sake, 114, 123; duck sex, role in, 117–18; individual, focus on, 115–16
aesthetics, 13–14, 18, 20, 113, 124; aesthetic intelligibility, 6–7; beauty, 81–82, 110; memory and testimony, 12; of mundane, 5; in nature, 110, 123; observation, 138; post-Kantian approach to, 22; and rhetoric, 148n15; sciences, relationship between, 18, 31–32; sensus communis, 11; subjective, 125; theory of evolution, 15
affect, 7–8, 12–13, 20–22, 25, 28–29, 44, 123, 139, 154n50; and beauty, 114, 140; and feeling, 31; paradigmatic, 141; punctive listening, 56; studies of, 148–49n18; theories of, 150n37, 153n37; whale affect, 97
affect theory, 150n37, 153n37
Africa, 54, 61–63, 72–73, 74–75, 78, 80
Alaska, 106
alogos, 124
Amazon, 15–16, 128
Amboseli National Park, 75
American Academy of Arts and Sciences, Committee on Space Efforts, 18
Amundson, Ron, 115–16, 165n30
animal behavior, 77–80
animal communication, 37–38, 56, 59, 73, 98
animal rhetorics, 148n5
anthropogenic noise, 83–85, 98
anthropomorphism, 14, 39, 77–78, 131; banning of, 79; beauty, 81; human exceptionalism, 79; as integral to felt objectivity, 60, 80; prescientific invention, 80; as problematic, 132; as product of anthropocentrism, 79; stigma of, 79
ants, 38
arachnophobia, 124
Aristotle, 7, 124, 148n11
artificial intelligence (AI), 160–61n63
Asia, 159n10
Avilés, Leticia, 90
Ayto, John, 92–93

Bacon, Francis, 24
Ball, Gregory F., 113, 165n14
Barad, Karen, 149–50n26, 152n10, 155n61
Barthes, Roland: *pheno-song* and *genos-song*, 158n72; photographs, 27–28; punctum, 14, 19, 27–28, 48, 154n50, 155n51, 158n72, 161n73; semiotics of, 28; studium, 27, 154n48
Bar-Yam, Yaneer, 163n44
bats, 36
Beautiful Brain, The (exhibition), 5
beauty, 6, 11, 14, 18, 29–30, 84, 86, 103, 113, 118, 121; aesthetics, 81, 110; and affect, 114, 140; anthropomorphism, 81; as arbitrary, 114–15; bearing witness, 55; beauty happens, 114–15; "being swallowed" by, 15; of birds, 124; of birds, and female choice, 111; in country music, 138; in dissonance, 139; in the everyday, 5; evolution, role in, 109–11; in evolutionary biology, 127; experience of, 7; and feeling, 31; and fitness, 110, 112, 123; humility, 7–8; idea of, 3–4; as imperative to witness, 12; listening to, 3, 16, 82, 138; lived experience of nonhuman kin, 15–16; making kin, 107; meaning, 44; mega-cognitive

beauty (cont.)
    abilities, 126; in nature, 123, 128, 137, 139–40; in nature vs. artwork, 22; in noise, 15; objectivity, 55, 125; observable world, tangible effect on, 110; open semiosis, 110, 114–15; painful side of, 101; response to, 99; rhetorical turn to, 19; in science, 90, 109, 134–36, 138, 141; in songs, 95; of sounds, 99
Becoming Wild (Safina), 6
bee-eaters, 72
beetles, 50
Berlin, James A., 152n6
Bermuda, 38
Berthoff, Ann E., 152n6
Bickford, Susan, 148n17
big listening, 95
bioacoustics, 3, 15, 37–39, 43, 61, 87; monitoring grid, 73–75
biosemiotics studies, 157–58n69
birds, 4, 18–19, 71–72, 82, 93, 109–10, 111, 114, 122–24; genitalia, 118; mating rituals, 118–21. See also individual birds
birdsongs, 14, 156n20
Borgia, Gerald, 113, 165n14
Bostwick, Kimberly, 112, 114
Boyle, Casey, 11, 65, 150n36
Braiding Sweetness (Kimmerer), 106
Bream, Shannon, 122
Brennan, Patricia, 15, 110, 119–20, 123, 134, 136, 139–40; anthropomorphic tendencies of, 131–32; "bizarre," observation of, 118, 121–22, 129–30; comportment, 128; duck genitalia, study of, 111, 117–18, 122, 127, 129, 132; felt objectivity of, 133; and Katy Payne, 121–22; microaggressions toward, 167n88; punctive listening, 122, 129–30; surname of, 133; "swallowed" by nature, 128–30; "Termite Lab," 130
brown, adrienne maree, 93
Buehl, Jonathan, 149n19
Burke, Kenneth, 17–18, 151nn1–2, 151nn5–6, 155n58, 161n81
Butler, Octavia, 93
butterflies, 109

Camera Lucida (Barthes), 19, 27
Campt, Tina: felt sound, 26
carbon footprint, 85
Carlson, Tucker, 122
Carola's parotia, 112, 120
Cassirer, Ernst, 152n6
Central Africa, 159n10
Central America, 112
Ceraso, Steph, 149n24
"Cetus: Life After Life" (song), 103
Changing Role of the Embryo in Evolutionary Thought, The (Amundson), 115

Charmaz, Kathy, 150–51n38
Chuh, Kandice, 11
Clary-Lemon, Jennifer, 147n2
climate change, 1, 70, 85–86, 91–92, 163n37; as "time of urgency," 90
Cline, Patsy, 138
Colombia, 118
colonialism: scientific research, 67; settler colonialism, 65–67
comportment, 86, 94; nature's punctum, 50; punctive listening, 50–52, 128, 139–40
Conference on Science and the Modern World View, 18
conservation, 75, 80; and listening, 74–75
Convention on International Trade in Endangered Species, 70
Cooper, Marilyn M., 148n5
Cornell University, 104–5; Cornell Lab of Ornithology, 72; "elephant dictionary," 73; Elephant Listening Project, 15, 59–64, 70–75, 77, 160–61n63; Herbert F. Johnson Museum of Art, 83, 103; K. Lisa Yang Center for Conservation Bioacoustics, 61–62; as "Temple of Adaptation," 121
Costa Rica, 86–87, 89, 98, 106–7, 118
Countless Aspects of Beauty, The (exhibition), 5
COVID-19 pandemic, 89; "anthropause" of, 15, 98
Cowley, Malcolm, 17
Coyne, Jerry, 113, 131, 165n14
Crick, Francis, 149n19
Critique of Pure Reason (Kant), 152n11

Damasio, Antonio, 150n37
Dancing with the Birds (documentary), 118–20
Danto, Arthur C., 6, 26, 114
Darwin, Charles, 5, 110, 113, 149n19; theory of evolution, 15
Daston, Lorraine, 149–50n26; naturalistic fallacy, 165n16
Davis, Diane, 47, 148n5, 151n1
deep listening, 102; punctive listening, akin to, 68
"Definition of Man" (Burke), 17–18
Dewey, John, 152n6, 153n19
diasporic listening, 138
diffraction, 26; as methodology, 20, 152n10
diffraction patterns, 21f
diffractive thinking, 20
disability studies, 164n68
dolphins, 4, 18–19, 36, 85–89, 98, 100, 129, 162n17; echolocation, use of, 94
Druschke, Caroline Gottschalk, 149n19

East Africa, 159n10
economic injustice, 95
Ecuador, 46, 87, 115
Eidsheim, Nina Sun, 49

INDEX

Einstein, Albert, 23
Eisner, Thomas, 50
elephants, 4, 18–19, 54, 59, 60–65, 62f, 72–73, 82, 159n10; anthropomorphizing, 78–79; in bais (forest clearings), 78; beauty of, as different kind, 81; call types, 61–62; "elephant dictionary," 73; gunshots, 74–75; individuality of, 77; infrasonic rumbles, 3, 15, 43, 50, 59, 61–62, 64, 73, 74, 75, 79, 80, 139, 161n65; ivory, 70; kinship, 81; "let's go" call, 62; listening through the ground, 76; poachers, 74–75; population of, 70; preservation, 80; punctive listening, 80; remembering, 63; "soap opera," like watching, 77; trunks, use of, 80
Ellison, Ralph, 17
emergence, 163n44, 165n29; "gere," 93; synoptic nature of, 94
emergent listening, 15, 85, 101, 138; as punctive listening, 94, 97
empiricism, 22–23, 35, 131
enclave theory, 95
epigenetics, 155n62
evo-devo (evolutionary developmental biology), 115–16
evolution: adaptation narrative of, 116, 120; agency of individuals, 116; female agency in, 111
evolutionary biology, 3, 15, 108, 110, 112–16; beauty in, 111, 127; female, as sexual subject, 111
evolutionary ecology, 47
evolutionary emergence, 86
Evolution of Beauty, The (Prum), 5, 113–15, 117, 126
Extavour, Cassandra, 23
extractive listening, 15; as defined, 60; as "hungry listening," 65–71, 85, 132, 140; settler colonialism, shaped by, 66; whiteness, 65–66

Fathom (documentary), 96
Feeling and Form (Langer), 20–21
felt life, 26, 28–32, 45, 48, 55, 154n43; in natural environments, 34–35; vivid values, 27
felt objectivity, 133
felt sound, 26, 45
female subjectivity, 15–16
First Nations communities, 66
Florida International University, 87
Fournet, Michelle, 15, 41, 85–86, 98, 102, 106, 139–41; big listening, 95; listening, practice of, 97; whale calls, 95–96; whaleness, 105; "whup," 97
Frankel, Adam: "Marine Heat Blob," 90–91
Frost, Robert, 108

Gabon, 72
Galápagos, 130
Galileo, Galilei, 134
Galison, Peter, 149–50n26
Gardner, Howard, 18

Garland, Ellen, 41, 96
gender, 133; gender-in-the-making, 149n25
ghost forest, 1–2
Gigante, Maria E., 149n19
Global Conservation, 70
Global South, 153n35
Goodall, Jane, 71
Gould, Stephen Jay, 121
Graham, S. Scott, 148–49n18
great tinamou, 118
Greece, 5, 76
Greenpeace, 35
Gross, Alan G., 149n19
Gross, Daniel M., 8
Grosz, Elizabeth, 154n43

Habronattus jumping spider, 16, 18–19, 124–25, 125f, 135; route memory, 126
Hannity, Sean, 122
Haraway, Donna, 11–12, 147n3, 149n25, 152n10; objectivity, as "view from nowhere," 10, 55, 149–50n26
Harding, Sandra, 149–50n26
Harmon, Joseph, 149n19
Harvard University Press, 152n16
Hawaii, 90, 156n18
Hawhee, Debra, 7, 92, 124, 132, 148n11, 148–49n18, 151n1
hearing, 96; felt life, 45; and listening, 10, 40, 51–52, 68, 76–78; as panoramic, 44; prenatal, 49; and sound, 45, 75–77; as symbolic, 161n73
Hedwig, Daniela, 15, 59, 77–80, 139
Hegel, G. W. F., 20
Hellenic National Archaeological Museum, Athens, 5
Herman, Louis M., 41
Hoffman, Julian, 63
Hoffmeyer, Jesper, 157–58n69
Horowitz, Alexandra, 51
Horowitz, Seth, 44–45
How Forests Think (Kohn), 48
How the Songs Come Down (Revard), 109
Hoy, Ron, 16, 134–35, 137, 139; jumping spiders, study of, 124–26; as "meta," 138; as observer, 138
Hsu, V. Jo, 164n68; diasporic listening, 138
humility, 7–8, 45, 51, 71, 128; punctive listening, 52
hummingbirds, 69, 71, 115
humpback whales, 101–3, 141
humpback whale songs, 3, 14, 33–38, 43, 46, 50–51, 54, 57, 58, 83–85, 99, 103, 139, 156n18; sexual selection, 41; sounds of, 4–5; whale calls, 95–98; "whup," 41. *See also* whale songs
"hungry listening," 65–71, 85, 132, 140
Hungry Listening (Robinson), 64–66, 68
Husserl, Edmund, 20

Immense World, An (Yong), 5–6
India, 159n10
Indigenous languages: verbs, emphasis on, 69
Indigenous peoples, 65–69
Indigenous sound studies, 60
Indigenous studies, 147n3
International Whaling Commission, 54
intuition, 22, 24, 31, 47, 57, 78–80, 85–86, 100, 139–40, 155n58; bodily technique, 96–97; as punctive listening, 106; and science, 81, 106; as way of knowing, 29–30, 32
invention, 67, 92–93, 133; aesthetics of, 16; community-led, 140; creative moments of, 19; logic, 152n11; multimodal milieu of, 12; scientific, 4. *See also* prescientific invention; rhetorical invention

Johnson, Mark, 12, 148n15, 153n19
Jones, Natasha, 151n41

Kanien'kehá:ka peoples, 67
Kant, Immanuel, 6, 20, 22, 28, 152n11
Keeling, Diane, 148n5
Keller, Evelyn Fox, 13
Kennedy, George, 148n5
Kenya, 61, 72
Kepes, György, 18
Kimmerer, Robin Wall, 60, 69, 85, 101, 106, 135–36, 147n3, 164n73; ethical revolution, and language revolution, 68; kinning, 76, 147n3; naming, 80
kin, 77–78, 94, 101, 107, 141; concept of, 76; listening, 67; making kin, 76, 79, 138–39, 147n3; nonhuman, 3, 109; nonhuman and human, 7, 60, 86; notions of, 147n3; punctive listening, 59–60
king bird of paradise, 119*f*, 120*f*
King Solomon's Ring (Lorenz), 161n81
kinning, 77, 89, 147n3; in prescientific invention, 77
kinship, 76; elephants, 81
Kohn, Eduardo, 53, 107, 123–24, 157–58n69; open semiosis, 46–49; regrounding, 47; symbolic thought, 46–47
Kristeva, Julia, 158n72
Kuhn, Thomas, 14, 20; intuition, 30; paradigms, 141

Langbauer, William, Jr., 61
Langer, Susanne K., 14, 17, 19, 35, 40, 44, 46, 49, 59, 77, 79, 109, 123–24, 135, 151n2, 151n5, 152n9, 152n11, 153n19, 153n35, 153n37, 154n48, 155n62, 157n48, 161n81; aesthesis, theory of, 27; aesthetics and sciences, relationship between, 18; affective symbols, 166n55; ambidexterity of thought, 20; artwork, subjective experiences of, 21; and beauty, 29–31; evolutionary theory of human mind, 22; feeling, as action, 29; feeling, as part of scientific meaning making, 31; felt life, 26–29, 34, 45, 48, 55, 154n43; humans, as feeling beings, 26; "idols of the laboratory," 24, 108, 153n28; intuition, 155n58; listening, 23; logical thought, expansion of, 32; as "mother of contemporary rhetorical theory," 152n6; music, and immediacy of feeling, 23; music composition, 24; prescientific invention, 24, 26, 29–32, 34, 45, 79, 100, 153n27; "semblance of a thing," 154n42; symbolism, theory of, 18, 21–23, 25–26; visual art, 77
Levine, George, 114
Lewandowski, Annie, 42*f*, 57, 85, 102–3, 105–6; call to beauty, 84
Leys, Ruth, 150n37
Liboiron, Max, 66–68, 159n3
Lilly, John, 36
Limón, Ada, 57
listening, 23, 45, 50, 54, 56–57, 59–60, 66–67, 81, 94, 96, 106, 109, 136, 141; activism, as form of, 86; to beauty, 3, 16, 82, 138; comportment, 51; and conservation, 74; as feeling what you hear, 95; as full-bodied experience, 49; and hearing, 10, 40, 51–52, 68, 76–78; hungrily, 15; intuitively, 97; kin, 67; as listening through, 67; as more verb than noun, 69; as multisensory act, 149n24; nature's punctum, 44, 48, 126; as "other side of language," 8; pain of, 15, 98–101; as quotidian element of rhetorical invention, 9; and resonance, 75–76; sense of time, shifting of, 69; and sound, 68–69; unknowing, possibility of, 138–39; in "wild places," 69; witnessing, as central to, 55, 65–66, 140. *See also* big listening; deep listening; diasporic listening; emergent listening; extractive listening; hungry listening; punctive listening; rhetorical listening
lived experience, 2, 13, 16, 43, 86, 110
logos, 124
Lorenz, Konrad, 79, 161n81
Louisiana, 1, 137
Lynn, Loretta, 137–38
Lyon, Arabella, 152n6

macaws, 6
manakins, 110–12, 124, 134
Mandoki, Katya, 7
marine ecology, 3
Massachusetts Institute of Technology (MIT), 18
Massumi, Brian, 154n42; emotion and affect, distinction between, 150n37
May-Collado, Laura J., 15, 85–90, 94–96, 98–101, 105, 164n68, 165n29; dolphins, study of, 162n17; phylogenetic approach, 139
McClintock, Barbara, 13, 23
McDonald, Kyle, 84
McQuay, Bill, 44–45, 51–52
McVay, Scott, 35–36, 40–41, 156n20

INDEX

Melville, Herman, 33, 36, 83, 109, 155n2, 159n4
methexis, 76–77, 80
metonymy, 155n51
Miller, Carolyn, 148n5
mimesis, 76–77
Mind: An Essay on Human Feeling (Langer), 17, 27
Moby-Dick (Melville), 36, 39, 83, 85, 131, 155n2, 159n4; "Saturn's grey chaos," 33
Morton, Timothy, 166n72
Moss, Cynthia, 61
Muckelbauer, John, 153n19

Namibia, 61, 159n10
naming, 80
Nancy, Jean-Luc, 76, 97
National Oceanic and Atmospheric Administration, 130
National Science Foundation (NSF), 111, 118, 122
nature's punctum, 4, 14, 28–29, 32, 34, 48, 86, 97–98, 141; comportment, 50; listening to, 44, 126
New Rhetoric, 152n6
Newton, Isaac, 149n19
Nezhukumatathil, Aimee, 33, 40
Nocek, Adam, 31–32, 155n62
noise, 86, 98; dolphin and whale communication, and human noise, 88–89; soundscapes, interfering with, 102
nonhuman, 35, 52, 60, 67, 98, 115–16, 134, 166n55; nonhuman animals, 11, 47, 54, 79, 111–13, 124, 132; nonhuman communication, 3, 46; nonhuman kin, 2–3, 5–12, 15–16, 31, 46, 53, 86, 109, 147n3; nonhuman ways of knowing, 14, 18, 34, 124, 148n5
Nouabalé-Ndoki National Park, 73

Oak Grove, LA, 1–2, 137, 147n4
Obama, Barack, 122
objectivity, 4, 14, 16, 25, 30–31, 43–44, 56, 77–79, 133–34, 149–50n26; beauty, 55, 125; felt objectivity, 80; prescientific invention, 80–81; as "view from nowhere," 10
observation, 9, 11, 23, 27, 50–51, 79, 149n24; as aesthetic act, 138; direct, 74; objective, 31; objectivity, 55; simple, 35, 44–45, 48, 54–55, 60, 68; visual, 12, 140; witnessing, 52–55
Ochoa Gauthier, Ana Maria, 161n74
Okun, Tema, 91–92
Oliver, Kelly, 52
Oliveros, Pauline, 68; deep listening, 102
On Earth We're Briefly Gorgeous (Vuong), 5
ontogenetics, 115–16, 165n29
Ott, Brian, 150n37
owls, 36

Panama, 72, 88
Papua New Guinea, 118–19

paradigm, 8–9; adaptation, 112; Kuhn's, 141; paradigmatic effect, 141; rhetorical way of being in the field, 141; scientific, 141, 149–50n26; shift in, 61
Parrish, Alex C., 47, 157n59, 157–58n69
Patagonia, 156n18
Payne, Katy, 5, 9, 14–16, 36–38, 40–43, 42f, 48, 59, 60–63, 62f, 64–65, 70–72, 77, 81, 85–86, 127, 132, 135–36, 139, 157n48, 158n96; American education system, and haste, 45; bearing witness, 54–55; comportment, 50–52, 128; deep listening, 35; discovery and revelation, distinguishing between, 53; elephant dictionary, 73, 80; elephants, studying of, 59–62, 64–66, 159n10, 161n65; experience, informed by, 55–56; felt life, in natural environments, 34–35; felt objectivity of, 55; humility, 128; as inductive, 45; influence of, 35; legacy of, 57; as musician first, 51; objective knowledge, 55; objectivity of, 55–56; as observer, 45, 50, 55, 131; open semiosis, 49; punctive listening, 35, 52–53, 55–56, 68, 122, 141; radical subjectivity, 56; rhetoric, 44; simple observation, practice of, 60, 68; as steward, 54–55; subjectivities across species, 56; subjectivity, 55–56; trans-species encounters, 65; "way of being," 121; way of listening, 45–46, 49–50
Payne, Roger, 33–41, 48, 54, 58, 83, 96, 130–31, 156n20
Peirce, Charles Sanders, 46, 153n19, 157n59, 161n73
Peters, John Durham, 9, 12, 92, 107, 135, 161n73, 166n55
Philosophy in a New Key (Langer), 17, 22, 152n16
phylogenetics, 115, 139, 165n29
physics envy, 24, 108–9
Pink, Sarah: sensory ethnography, 13
Pippin, Robert B., 6–7
poachers, 74–75
Polanyi, Michael: tacit dimension, 153n27
polyandry, 72
Poole, Joyce, 61; translation, 73
positivistic science, 19, 22
posthuman, 46, 112, 124, 127, 133, 148–49n18
posthumanism, 166n72
prescientific invention, 24–26, 29–32, 34, 45, 77, 79–80, 81, 98, 100, 141, 153n27; anthropomorphism, 79; objectivity, importance of, 80. *See also* invention
presence, 72–73; being in touch, 73
preservation, 80, 81
Propen, Amy, 148–49n18
Prum, Richard O., 5, 15–16, 124, 128, 134–35, 139–40, 164n68; adaptationists, argument with, 110–13; aesthetic evolution, 110–11, 114, 116, 118, 122–23, 126, 133, 165n30; as bird watcher, 110, 127; felt objectivity of, 133; as feminist, 111, 113, 165n14; first-person experience, 127; and individual, 115–17; as outsider, 115; physics envy,

Prum, Richard O (*cont.*)
    108–9; plumage of dinosaurs, 123; speculation, as "idea generation," 132; speculative science, 131, 133; tropical birds, plumage of, 110
public sphere, 5
punctive listening, 14, 35, 44, 48, 53–55, 60, 70, 77, 86, 101, 114–15, 122, 126, 129–30, 138, 141, 166n55; affect, 56; attention, 51, 102; as bearing witness, 76; comportment, 50–52, 128, 139–40; decentering human knowledge, 127; deep listening, akin to, 68; elephants, 80; as emergent listening, 85, 94, 97; empirical inquiry, method of, 44; as felt objectivity, 56; female subjects, giving voice to, 139–40; humility, necessity of, 52; intuition, 106; kin, 60; objectivity, 44; paradox of, 81; phylogenetic approach, 139; semiosis, 48; as witnessing, 76, 131
punctum, 14, 27–28, 48, 76, 155n51, 158n72, 160n37, 161n73; definition of, 19; etymology of, 154n50

race, 57, 133, 154n48
Ratcliffe, Krista: rhetorical listening, 149n20
Rendell, Luke, 40–41
Republic of Congo, 73–74
resonance, 77; and listening, 75–76
Revard, Carter, 58–59, 65, 69, 108–10, 159n2
Revard, Lawrence, 159n2
rhetoric, 2, 14, 17–18, 44, 151n1, 153n19; and aesthetics, 148n15
rhetorical invention, 9–10, 24, 26, 70, 148n17; wordlessness of thought, 8. *See also* invention
rhetorical listening, 149n20
rhetorical theory, 148–49n18
rhetorics of science, 4, 22–23, 31, 60, 149n19
Richards, I. A., 152n6
Rickert, Thomas, 148–49n18, 157–58n69
Rivers, Nathaniel, 85
Robinson, Dylan, 60, 65–70, 74
Rodriguez, Patricia, 133
Rogers, Fred, 5
Rose, Deborah Bird, 76, 147n3
Rothenberg, David, 35–36, 38–39, 44
Rowland, Liz, 70–71, 74
Rubin, Amy, 84
Runa people, 46, 48
Runia, Eelco, 73, 155n51
Russell, Bertrand, 153n19

Safina, Carl, 6, 75, 76, 79
Sakakibara, Chie, 52
Schaffer, Simon, 53, 149n25
science studies, 14, 18
scientific method, 9; witnesses, requiring of, 10
Scott, J. Blake, 149n19
Sedgwick, Eve Kosofsky, 150n37
Seeger, Pete, 35, 44

Seeley, Thomas, 50
semiosis, 46; open, 47–49, 56–57, 110, 114–15, 131
semiotics, 27–28, 55, 115, 123–24, 127
Sense of Urgency, A (Hawhee), 92
sensory ethnography, 13
settler colonialism, 65–67
Shapin, Steven, 53, 149n25
Sharpe, Christina, 7, 154n48
Sheffer, Henry M., 152n11
signification, 27
Siren—Listening to Another Species on Earth (Lewandowski), 83–84, 84f; visitors, attention span of, 102–5
Skjálfandi Bay, Iceland, 98
sociality, 123
song, 12, 30, 59, 69, 114; agency, amplifying of, 159n2; beauty in, 95; birdsongs, 14, 156n20; as genesis and survival, 159n2; *geno-song*, 158n72; and language, 47; *pheno-song*, 158n72; "siren song," 103; and whales, 33–34, 36–37, 39–42, 54, 83, 89, 96, 103–6; as written genetically, 58. *See also* humpback whale songs
Songs of the Humpback Whale (Payne, Payne, and McVay), 35–36, 41, 54, 83
"Songs of Humpback Whales" (R. Payne), 36, 37f
"Songs of the Wine-Throated Hummingbird" (Revard), 58, 69, 159n2
soundscapes, 86–87, 89, 102; as shared, 85
sound studies, whiteness of, 65–66
Spain, 5
Structure of Scientific Revolutions, The (Kuhn), 20
subjectivity, 30, 55, 127, 132, 134; female, 15–16; individual, 116–17, 154n88; radical, 56; in scientific practice, 31
symbolism, 18, 23, 25; affective, 21–22, 141; discursive, 22, 26, 39, 47; non-discursive, 22; presentational, 22; symbolic thought, 46–47
symbols, 152n18; icons and indexes, 46–48; as kinds of signs, 46

TallBear, Kim, 68
theory of evolution, and aesthetics, 15
Thomas, Elizabeth M., 61
throat singing, 66
toadfish, 101
trans-species, 64, 65, 87, 139
Tubb, Ernest, 138
"Tuft of Flowers, The" (Frost), 108–9
Tuhiwai Smith, Linda, 67

United States, 20, 87, 90–91

van der Tuin, Iris: diffractive thinking, 20
"view from nowhere," 10, 55, 140, 149–50n26
Vivian, Bradford, 92
Vuong, Ocean, 5

Waddington, C. H., 18
Walcott, Charles, 50–51
Walton, Rebecca, 151n41
Wartzok, Douglas, 87
Washington Park Zoo, 43, 60–62
Waterman, Ellen, 68
Watlington, Frank, 36, 156n18
Watson, James, 149n19
West Africa, 159n10
Westling, Louise, 148n5
whales, 4–5, 18–19, 36, 54, 70, 74, 83–85, 87, 89–91, 130–31, 141; common ancestry with, 33, 38; as composers, 34, 40–41, 50; whale calls, 95–98, 104–7, 109; whale-watching trips, 98, 106
whaling industry, 36
whaleness, 105
whale songs, 42–43, 47–48, 83, 89, 98, 103–4, 134; anthropomorphizing of, 39–41; emotional reactions to, 38–39; human identification with, 39; mixtapes, 57; patterns of, 37; spectrograms, 36–37, 44; structure of, 34; as Ur-music, 39. *See also* humpback whale songs
whale sounds, 96
Whitehead, Alfred North, 14, 17, 20, 24–25, 152n11, 152n18, 155n51, 157n48; abstraction, 154n41
Whitehead, Hal, 40–41, 46
whiteness, 133; extractive listening, 65–66; of sound studies, 65–66
white settlers, 147n2

white supremacy, "sense of urgency," 91–92
whooping cranes, 2
Williams, Brooke, 79
Williams, William Carlos, 17
Wilson, E. O., 38
Wilson, Ken: enclave theory, 95
witnessing, 1, 64–65, 97, 102, 106, 109; bearing witness, 51, 53–55, 76, 140; as beauty, 55; emergency, 92, 94; listening, 55, 65–66, 140; methods of, 12–13; "modest witness," 10–12; punctive listening, 76, 131; relationality, 10; scientific witness, 52–53; sensing a scene, 68; as visual or discursive act, 53
Wong, Deborah, 68
Won't You Be My Neighbor? (film), 5
world making, 76
World of Wonders (Nezhukumatathil), 33
World Wildlife Fund (WWF), 70
Wrege, Peter, 15, 59, 63–64, 70–75, 77–78, 81–82, 139; as data driven, 71; geodome, 70–71; workaholic tendencies, 70–71

Xanthopoulos, Drew, 96
xwélalà:m, 67
xwélmexw, 65

Yale University, 110
Yong, Ed, 5–6

Zimbabwe, 61

www.ingramcontent.com/pod-product-compliance
Lightning Source LLC
Chambersburg PA
CBHW022011290426
44109CB00015B/1142